£1-80 £1.20 Yeh!

LIVING ON A LITTLE LAND

BY THE SAME AUTHOR

The Restless Generation (1972)
Politics by Pressure (1974)
The Survivalists (Eyre Methuen, 1975)
Living Better on Less (Turnstone, 1977)

To John and Dammy,

and all others with
the same dream -
and an equal determination
to live it.

PATRICK RIVERS

Living on a Little Land

Illustrated by

SALLY SEYMOUR

Foreword by John Seymour

TURNSTONE BOOKS

Published by
Turnstone Books
37 Upper Addison Gardens
London W14 8AJ

Hardcover ISBN 0 85500 086 4
Paperback ISBN 0 85500 087 2

First published 1978

Typeset in 10/11½ Plantin by
Saildean Limited, Kingston
Printed by Biddles Limited
Guildford

Contents

Acknowledgements

Anyone who writes about self-reliance without having practised it risks telling people to 'do as I say, not as I do!' Shirley and I have done it, but it would not have been possible without help from others, firstly in our venture of living on a little land and secondly in writing this book about the continuing experience. And so I am happy specifically to record my thanks to at least a few who unstintingly helped the two activities in undeserved and unexpected ways: Charles, Dammy, Donald, Eduardo, Elizabeth, Gordon, Harry, Jack, Jim, John, Julie, Keith, Ken, Laurie, Lil, Martin, Masie, 'P', Pat, Richard, Robin, Roger and Steve. To these I must add all the 'Wwoofers', the others who worked with us for love, and of course my wife, Shirley, who did all you might expect – and immeasurably more.

All the events described are substantially true, though some details have been changed to avoid possible embarrassment. For their protection and my own, I have to add that none of the characters resemble people alive or dead.

Foreword

JOHN SEYMOUR

I have for a long time admired Patrick Rivers as a writer and liked him as a particularly intelligent, honest, and gentle man, but it was not until very recently that I visited him in his home. I went there expecting the worst. I have visited so many of the 'new countrymen' in their 'self sufficient holdings' and have so often been disappointed. Self sufficient the holdings so often are not: they tend rather to be a terrible mess. The owners always try to do too much rushing, from job to job and never actually finishing anything; finally, they very often, give the whole thing up in despair.

But of course I forgot that Patrick and his equally competent wife Shirley had been commercial farmers both in Britain and in Australia, and had spent long enough in the country anyway to have acquired something of the countryman's practical aptitude: a thing so often lacking in the city intellectual trying to turn countryman. To visit 'Field Gate' is a most rewarding experience. Here, on a few acres of nearly impossible land, (so steep that even the goats look giddy) these two people, helped quite a lot by sundry volunteers, have developed what seemed to me to be well on the way to being a model self sufficient smallholding. They really are producing a great deal of their own very good food: all their vegetables and potatoes, all milk and dairy products (their soft cheese was delicious), all their eggs, and a considerable surplus of food to sell or give away to other people. As they are vegetarians, the natural increase of their birds and animals that they do not need for replacements is exported off the holding to feed the rest of the world. They have raised the productiveness of this piece of land - so 'marginal' that no commercial farmer would consider it for anything but the roughest of rough grazing - very high indeed. One is made to realize, by seeing holdings like this, that the British Isles could very easily be self sufficient in food - and have enough to send away to other countries too.

A thing that we must always remember when comparing husbandry of this sort with orthodox commercial holdings is *the input/output factor.* That is the relationship between the true wealth that is put into the holding and the true wealth that comes out of it. Anyone, given an acreage of easily worked level land in England, and unlimited, cheap (government subsidised) chemical fertilizers, pesticides, herbicides, fungicides and other poisons, can produce an awful lot of food per acre. There really isn't anything very difficult about it. But to take land from which nobody else could make a living and - with no input of oil-derived chemicals and a very small input of power - produce good food, is very difficult indeed; like many difficult things, though, it is very worth doing. Probably the most worthwhile thing that any of us can do today.

At present our farmers and market gardeners are simply buying fertility and some freedom from disease from the chemical factories. And it is as well to remember that nearly *all* of their chemical and fuel inputs come from the oil wells. The day is going to arrive when these inputs will no longer be available, or at least will be very, very expensive. When it does come there could well be a food crisis in this world, of dangerous proportions. It would be admitted by any competent agriculturalist, no matter how much in favour of chemical farming he was, that British agriculture has got to the stage at which production per acre would fall abysmally without chemical inputs. Animals have been divorced from the land, with the consequence that the benign interaction between the two Kingdoms - vegetable and animal - has been destroyed. When we see the Rivers' mild-eyed heifer and their three idiosyncratic goats (one of them fell in love with my small son - a female goat I am glad to say) and when one sees the large and growing pile of honest muck outside their cowshed, and the health of their crops, one realizes that these different factors are not unrelated. Human toil - the hoe replacing the selective weed sprays (herbicides), the manure barrow the fertilizer bag, good husbandry the huge battery of chemical poisons now at the disposal of the agribusinessman - is turning this hillside into a Garden of Eden.

Man, we are told, was put into the Garden to 'dress it and keep it'. Not to rape and poison it, extracting the last bit of profit before it turns into a desert. It is inspiring to see people such as the Rivers - husbandmen in the real sense of the word - toiling as Nature meant us to toil - to secure that proper balance between the soil, the climate, and all forms of living things, including Man himself, which produces the only true health - that is wholeness - the only real commonwealth that can possibly survive on this planet once the great bonanzas of free power have run out and much of the land surface has been exhausted by extractive agriculture.

But this is not a polemical book. Not that the author is lacking in fire in his soul or a fierce concern over the future of life on this planet: it is just not that sort of book. It is a human, witty, delightful account of a family that has accepted a challenge, and with labour, love and devotion, measure well up to it. Besides inspiring us, which it does, it will amuse and charm. In fact it is a damned good story and well told.

CHAPTER ONE

The Reality

To live each day surrounded by beauty rather than bustle, to breathe sweet-scented air in place of fumes, eating food that tastes as it used to do, with an appetite sharpened by outdoor work; in short, to provide most of your needs with your own hands from your own land. This is a hope embraced by millions.

It is a tantalising dream, for thousands have already made it reality: they are out there working their own land; free from the treadmill of compulsively earning more and more to keep pace with their rising expectations and expenses, free from the slavery of boring and meaningless work. They are out there living a vastly more enjoyable and fulfilling life, in which no-one tells them what to do, with far more control over their destinies, and on a fraction of their previous city earnings. Such a life grows more alluring as city life grows grimmer, for we know now: it can be lived! This message, coming from those who have freed themselves, is not going unheeded. They are being joined daily. Women are realising that a suburban box of a house and a pocket handkerchief garden can become a prison of boredom and isolation. Men are seeing that worrying and working all day at a desk, eating, drinking and smoking too much all lead to the inevitable coronary and more. The dangers of eating food grown with poisons and processed with chemicals are stark. Motives for exodus abound.

The country life is not simple however; to live it happily you need intelligence, a zest for work, determination, some money and a sense of humour, as well as a sensitive feeling for nature. Even with all these you may still fall into one of many traps. To others you may *appear* successful, with all your aims realised, yet not be truly happy. Because we are all naturally reluctant to admit our shortcomings, little is heard of those who try 'the good life', only to return to the city, taking with them priceless lessons they might have passed on to others; nor do we learn much from those who hang on grimly, for they tend to pin on brave smiles and bury their unhappy experiences deep.

So what of the many articles and books on the subject, already written by truly successful self-supporters? Surely they can tell the newcomer all he should know, or provide a ready checklist for those who have made the break? Not necessarily. Too often they deal with the purely practical, skimming over information already covered more deeply in books on specific subjects. They give scant attention to the crucial subject of money, both capital and income. What they most neglect is the essential skill of organising; they fail to put together the many different parts of country life in such a way as to make a working model; so that to rely on them for guidance would be about as futile as trying to create a human being by reading all the books on anatomy. They ignore the hard economic fact that even in the highly unlikely event of your growing *all* your own food and fuel, as well as making clothes, soap, candles and so on, you would still need an income - which must usually be earned, often just where money is scarce. They do not cover the subject as a whole. They are rarely helpful about coping with the problems of adjusting to country life, and they say too little about the frame of mind in which the would-be self-supporter should approach this new way of life - the feelings and personal qualities which can make the difference between success and failure, between happiness and disillusion. You can see why. It is far simpler to milk a goat and to sow a row of carrots than to learn to live with the elements, adjust to loneliness or feel at home among neighbours with different values from your own, with entrenched customs foreign to you. So, if we air these subjects, pointing out the mistakes of others with experience, then we shall help those considering this way of life, just as we shall help those who have already adopted it.

Rural self-reliance can offer unparallelled joy and deep satisfaction to those who prepare themselves for it. However, many fail because they were never suited to its demands, or they chose it for the wrong reasons, or neglected to do their homework beforehand, or made more mistakes in the doing of it than their resources of money, energy or determination could withstand. Most of this sadness could have been avoided with good advice given soon enough. The time is right for trying to put this new movement of returning to the land into perspective, examining its many parts and arranging them in such order that they add up to a meaningful philosophical statement, rather than a piecemeal escapist fantasy. They need to, for cocking a snook at the rest of the world is hardly a *raison d'etre*. You need a better aim to keep you going through the years ahead.

Coming down to earth

If the thousands now enjoying rural self-reliance are to be joined by more - whether one reader of this book or an exodus - they ought to know what they are in for. Illusions about country living grow as thick as nettles round a dung heap. Word has got around that a self-supporter's days are spent in unruffled calm. All is beauty and order, mornings of promise, sweet-scented evenings, mugs of home-made cider, and hay safely in the stack. The sun-tanned husband quietly goes about his manly work in the fields, joined in lighter tasks by his rosy-cheeked wife, fresh from butter-making in the scrubbed dairy. They have ample time to savour the view, the sounds of the birds, the scent of wild flowers. Goats and cows graze peacefully in gentle sunshine. Neighbours drop in, children play happily. At the end of the day they reminisce, make music or love until untroubled sleep overtakes them. Each day they retire early and wake refreshed at the birds' dawn chorus. Each day passes in gentle harmony with nature, the willing ally in their venture. Life in the country, however, is not like this. It is no more a true picture of life on our place than on anyone else's that I know. For our first few years it was jobs interrupted and left half-done, a back-log of work which never grew less, and constant vigilance for the next disaster. Our place was not the haven of unruffled calm we dreamed of in our city apartment. Let me paint a picture of reality which took shape - and still lingers.

Mornings I would try to spend writing, but our mail was heavy, and correspondence ate into the time available. Phone calls and visitors constantly interrupted. Afternoons I would work on the land, for if I tried writing I tended to doze off. I might settle down to servicing the rotovator, only to discover that the hens had got into the vegetable garden and I must fix the fence. Or perhaps Shirley and I had just got into a good rhythm hoeing when one of us would remember that we were almost out of wood - and it looked like rain.

For both of us, work indoors constantly competed with work outside. And usually the second won. Shirley would be affronted by the dust in the living room and she must decide whether to leave it yet another day or join me slashing bracken. On her way out she might remember as she passed through the kitchen that we would have no bread if she neglected to bake. But I would find perhaps that in my haste to get out to the bracken I had neglected to grind any flour and so she must knock up a quick soda loaf of left-over flour and oat flakes later. She would join me until it was time to feed the hens, chop the morning wood, bring in the goats, milk and feed them. The day passed quickly.

Evenings we would try to set aside for reading, talking, listening to music, writing personal letters - and just *being*. Too often we spent them instead planning the next day, the next month, discussing whether or not to buy this or that for the farm, or doing jobs for which there was no time during the day: grinding flour, baking bread, bottling cider, maybe sharpening tools for the next day's work.

Today our place is still untidy; things are tied together with baler twine and propped up with bits of old wood; there is chronically too much to do and we are always behind; jobs begun long ago have been left unfinished as others take priority. Perhaps we should not go to bed, I sometimes think, for we often wake up more tired than when we lie down - feeling as if we have someone else's bones in our bodies. All the same we work and move around at a fairly brisk pace. We concentrate hard on whatever we are doing. We need to be constantly vigilant, watchful for bugs with evil designs on our crops, checking our fences in case the goats seize an opportunity to prune the fruit trees, heedful of goat bells or their silence, noting the temperature of the greenhouse, scanning the skies for changes in the weather. Sounds, sights, smells and temperature send us continuous messages of reassurance or concern.

Sometimes we get a feeling we are being 'tested' - that we are given barely enough time after having dealt with one problem before the next one comes along. Difficulties seem meted out to us in doses large enough to break us if we failed to muster our last ounce of energy.

Our diary is revealing. Just before our third Christmas, during a spell of bad weather, we decided to decorate our bedroom at last. Gaily we began scraping excess plaster on our Attractive Exposed Beams - and found rampant woodworm. We checked our insurance and congratulated ourselves that it would take care of everything. Until the company surveyor climbed his patent ladder and flashed his torch into the roof area: pools of water lay on our clever ceiling insulation; green mould covered the rafters and trusses; he found everything except stalagmites and stalactites. He frowned, tapped The

Policy, and explained how we must insert an airbrick in the gable end and a vent in the trapdoor. Simple. Except for banging a way through the 18-inch thick wall on top of our tallest ladder and then crouching in a three-foot high dank-smelling loft to shift a hundredweight or so of masonry and make good; which an agile young man kindly did for us.

Soon after this came The Great Rain and, our morale repaired, we focussed our care again on the land. Here we found that a ditch dug in The Great Drought had burst and small rivers were running swiftly through our military rows of leeks, eroding our light soil. We dug and repaired it, then noticed what was happening to a nearby haystack. Here water had collected in puddles on the plastic cover and was trickling into the precious hay beneath. We removed the cover, took out spoiled hay, remade the stack into a handsome dome and replaced the cover, trying it down with bricks and twine. That night came The Great Gale which ripped the cover half off. I fetched more bricks and twine until it looked like a mammoth registered parcel. All *must* be well now we thought - until I noticed a small waterfall coming out of the bottom of the stack. We removed the cover again, carried all the good hay on pitchforks into our one dry shed, and discovered a brand new spring under the stack, gaily bubbling up among the last few inches of hay - now ruined.

The other kind of spring, we consoled ourselves, was on the way and as an act of faith we laid out neatly our early seed potatoes to sprout in an airy shed. Then, a few days later, a thought came to me: 'Shirley,' I said, 'I think I'll move those potatoes into the front porch. If not, mice might nibble them,' I went to the shed. I was wrong: they were not nibbled, they were nobbled; all except a couple of pounds of the biggest had gone. Rats, I learned, can happily cart off a small potato. And why should they care if seed cost 70p a pound?

These minor incidents added up to one of our sporadic bad patches. Life for us is not always as bleak of course, we enjoy rewards such as you cannot buy, as you will discover in the unfolding of this book. Nevertheless the going has been tough right through, and when it has all been too much, Shirley and I have shown the strain by being unnaturally polite to each other, and I have woken in the nights to worry. This being so it would be dishonest for me to portray 'the good life' as unbroken tranquility and bliss, even though many self-supporters may think us traitors to the cause that I now do so.

We know that the money and work which we have lavished on our place offer no guarantee at all that we can stay here. We must live with a hard truth: no matter how much we have become part of this place, we can stay for no longer than we can balance our books. Even though we might manage to eat free and collect our own fuel cheaply, there will still be bills to pay. The State offers no soft cushion for the likes of us if we should suffer crippling accident or long illness, or if my writing markets should dry up. Come the day when we are no longer fit for hard outdoor work or able to pull in a steady income, any cosy dream of 'self-sufficiency' will be over.

Simply because all has not always gone well for us we feel qualified to speak up. We have run the gauntlet. We have seen ourselves change: we questioned our comfortable city life-style: we exchanged a well-paid job for the uncertainty of a free-lance craft; put our apartment up for sale, searched for land and eventually bought eight derelict acres; we practically rebuilt the house, turning our enterprise into a working unit; and at the same time we made - for the second time in our lives - the profound adjustment to country living. We are not the same people any more.

We practically rebuilt the house

The rewards of change

We are better for the change. For one thing, we no longer divide our days into artificially rigid compartments labelled 'work' and 'leisure'. The two have blurred until the divisions have almost gone. To begin with, we are not at all sure any longer about the difference between the two. I had begun to think that work was essentially an activity you were paid for and reluctantly performed usually between set hours - say nine to five - between Monday and Friday; that it was *useful*; and that you went away from home to do it. Now I am not so sure. Most of my work had never actually *produced* anything, and when it did you could hardly call its end-product 'useful'. Now most of my work earns us no money; I do it at any hour of the day, and day of the week; much of it I enjoy more than my previous 'leisure'; all of it I do within the farm boundary, and often alongside Shirley or other people of my own choice. Moreover, its end-product is unquestionably useful. There is no set time for 'leisure'. Certainly there is time for doing what we want to do; but if what we want to do is as likely to be planting some crop as reading the paper or planting flowers, how can we split our lives in two? And even though it has become an imposed convention to do so, just how natural, we ask, is such a split life for any species? We have found that the way we live uses all of ourselves: our heads, hearts and bodies and this is surely a very healthy way.

This is not all, however. By stretching ourselves to our limits, mentally and physically, we have discovered new resources within ourselves and revived old ones which life in the city had suppressed. We are not so helpless. When we need warmth we no longer reach for a switch, we go out and saw some wood. When we want food we are far more likely to pick or dig it from the garden or take it from our ample storeroom than go to a shop. If we need exercise there is always work to be done. When we want stimulation it is all around us, for the country is an exciting, unpredictable place of limitless variety - with people to match. And when we need replenishment we rarely have need to travel; people travel *to us*, and some of the most beautiful scenery imaginable encompasses us. To live any other way has become unthinkable.

The truth of this came to us one evening, relaxing in front of the fire after a particularly hectic day. I turned to Shirley and asked: 'What do you miss in this, our new life? What did you enjoy in our city days which you miss - not just a little, I mean, but often and badly?'

I could tell by the time she took to answer that she was giving the question careful thought. Then she said: 'Concerts, art exhibitions, a good film, perhaps. But no, I don't miss them desperately, only sometimes, and something soon comes along to take their place.'

'A drink before our evening meal?' I suggested.

' Of course, ocasionally; but again, not very much. What I do miss sometimes is not having to *think*. In London I wasn't involved in your work and life for me was pretty routine. Now that we share the work I have to be much more involved in thinking ahead, holding so many different projects in my head at once, and all of them in varying stages of development. In the home there's the housework, cooking, cheesemaking, preserving, bulk-buying, catering for visitors and so on; and outside there's seed to order, rotations to follow, planting, harvesting, and of course the constant alertness anyone must have who keeps animals. But of course I feel much more alive than I did. And that's good. What about you?'

It was my turn to hesitate. Then: 'Sometimes when someone starts talking about the East, or a place I've enjoyed visiting, I get this urge to travel again. But when I think of the effort ... not really. A new camera perhaps, but I can manage with the one I've got; and when it finally gives up, I know Hal will lend me his ...'

And so we continued, conjuring up wants and then seeing them disappear like magicians' rabbits.

Presently I asked a different question: 'Imagine we had to leave here and go back to our old life in the city. What would you miss deeply?'

This time there was little hesitation on her part. 'Oh, the animals, the beauty of the place, growing our own food, the people we've come to know, seeing our plans take shape, working together out of doors, the wildlife ...'

I joined in, and the fire had died to embers before we were done. Yes, indeed: to live any other way had become unthinkable.

When Shirley and I planned our move we each held an image in our minds of the kind of place, the district and the tempo of our lives. We saw ourselves tucked away at the end of a long track, but our place is by a busy road; we thought we would be isolated, yet hardly a day goes by without someone dropping in - either from the neighbourhood or afar, friend or stranger - for word soon got around that we are up to something different; we believed we would soon have time to spare, but we are still looking forward to it, consoling ourselves that even if we have often been anxious, we have almost *never* been bored; we thought country life would soon be simpler, though only now is it shaping that way.

If you are planning a move, you too will build images of your new life, and I would hazard a guess that, unless you pick an outlandishly remote spot, loneliness will prove the least of your worries. To begin with, the unexpected variety of life is one feature of country life that you will soon encounter. You find that human beings, cats, dogs, pigeons and sparrows are not the sole inhabitants of this planet. Suddenly you are in good company. Each of your farm livestock has its own character, its own wants, which it will rapidly make known to you. You make their lives comfortable, they in turn give you their surplus produce of milk, eggs and honey. Apart from your own livestock you should have ample company in the creatures that come in from the wild. With an accommodating attitude of mind you can see that you share your land and your district with others and, in ways that may not be immediately apparent, you are mutually dependent.

Get to know these other creatures, for they will instruct you, entertain and comfort you. When bureaucrats hassle you, scientists frighten you and politicians exasperate you, the exquisite beauty and order that exist in the seeming turmoil and confusion of nature can free you from the tangle of circumstances in which we are caught, and to which we frequently attach undue significance. You are now in the real world of infinite mystery, delicate balance and limitless variety, so different from the fabricated environment of the city, where seeming order barely conceals chaos.

Simple living

Rural self-reliance *can* be simpler than city life, though it seldom is at first. Later, however, many of the new skills, disciplines and chores which have to be learned and remembered will become second nature. To begin with, there are rarely shops just round the corner. If you run out of food while cooking or forget to order all the equipment to start some project on time, you can have problems. You must either make do with a substitute, borrow from a neighbour if you can, or change your plans. It is much easier to buy bread than bake it. If you have been used to central heating, the right temperature is merely the turn of a switch or two; but if you are relying on wood you must first build a shelter for it, and figure out a way of carting it home; then find either suitable dead wood or a stand of live wood months before you need it; unless you have your own chain saw you will have to hire or borrow one, ensure that it is working and that you have plenty of fuel; then fell it, saw it, and cart it, stack it under cover until it is seasoned and dry, then saw it to the right length, and bring as much as you need indoors - making sure your chimney is swept, your stove is in good order, and you have plenty of dry kindling wood. Homemade bread, homemade beer or wine, your own honey ... everything you make yourself sounds like music. It promises a keener taste and extra nourishment. But to buy everything readymade from a shop seems to be so much easier.

I remember one of our visitors well. After a gentle walk to the end field and back he joined us in a meal of home-baked bread, freshly picked salad and goat's cheese. 'In the city now,' he observed, 'This would cost a packet. Lucky you with all this free food,' Shirley kicked me under the table and I stifled an imminent explosion, reminding him instead that before we could serve up our free cheese we had to clear an acre of bracken and brambles and sow pasture, make a ton of hay and feed it out all winter, build a goat house and find straw for bedding, roam the country to buy goats, take them in the van to Mrs. Jones's billy, drench them for worms, feed them up on crushed oats and barley, buy milking buckets, strainer and cheese press, get up in the night when they kidded, milk them daily - and then make the cheese! And I told a similar tale of the bread and the salad. One day, maybe in ten years' time, when we have amortised all our costs, our food may be free, but right now it certainly is not.

So why self-reliance? The point is that it frees you from the hidden price you are forced to pay to enjoy the excitement of living in a city: the price is mundane work, commuting, pollution and stress - especially the stress of work which troubles your conscience - and a host of other problems. Only when you balance the absence of these against the fuss of making and growing your own does the subtle simplicity of country living become apparent.

Country life today is satisfying, healthy and exciting in a host of ways which are apparent throughout this book. All the same, the qualities of the life are not obvious; if they were perhaps there would not be the present exodus. No neon signs advertise 'Badgers' Lair'; no high-decibel mood music draws your attention to the sunset; and the birds perform their dawn chorus at a most untimely hour. The trouble is that country attractions are just not stage-managed. The day is not neatly and clearly divided into work and leisure. The joys of country living are free, and if you have become used to *buying* your art and entertainment you are likely to pass over them. You have to discover them for yourself, and unless you have at least some of the qualities I have listed, the joys will escape you and you will yearn for your city entertainments as before — as like as not slumped before your TV or 'hi-fi' each evening, Sundays spent ritually poring over the newspaper sections. Unless you have some of these qualities, you will feel isolated, and you will take longer getting to know country people. Nor can you necessarily expect a sense of community with other self-supporters, for there are still relatively few and they are too scattered, too busy with their own thing.

Suppose you do miss the subtle enchantment of this way of life, and you find consolation in the knowledge that you are making your place run smoothly, growing your own food and becoming more self-reliant? Even if the happiness of living each moment of each day is denied you, can you not take heart from knowing that what you are doing has relevance to today's major problems? To *some* problems perhaps, but not all. Beware of complacency. Talk - as Shirley and I have done - to working people without money, trapped in a city environment, with neither useful skills nor friends in the country able or willing to rescue them. Sense as you talk, their utter helplessness, their frightening vulnerability in the event of economic breakdown. Listen to world problems from *their* point of view, and then decide for yourself just how much relevance your few acres and your life style have for them.

Motives for moving

Does the idea of living on a little land still appeal to you? Do you feel tempted to leave your city work and pleasures, your home and friends to 'have a go'? If so it would be well to cross-examine yourself first and check how realistic your motives are. Plenty of people have made the move, only to discover after their first summer, when the winds howl and everything turns to mud, that somehow they are not finding what they set out to find. Suppose we take a look at some of the more realistic motives for contemplating what could prove to be the most radical change to your life that you have ever made.

You look round you in your city environment and what do you see? Bricks and concrete, traffic, people, cats, dogs and - if you are lucky - a few oases of trees and grass set aside 'for enjoyment'. One day you begin thinking about your work and the way you live, and in time the sense of it steadily evaporates. Most people get no further than regrets, coupled with redoubled efforts at fleeting escape; a few however seek worthwhile alternatives, either where they are or somewhere else. Living on a little land is just one of many such possibilities. To escape from the 'rat race' then is a perfectly valid motive for moving, always provided that it is not the *only* one.

As Shirley and I planned our new life we found ourselves seeing the city far more critically. We began challenging practices we had previously accepted. This may happen to you. If it does, you may begrudge, as we did, the convention of living in two places - workplace and home - which may be miles apart. The cost in money and time can suddenly strike you as monstrous, and then the alternative of living and working in one place becomes increasingly attractive. You may find yourself also questioning how much money you *really* need to live on. Why toil so pointlessly at set times for more and more which buys you less and less genuine satisfaction? Why risk your health in doing so when it becomes patently clear that a life of too much stress, sitting and self-indulgence is a ticket to trouble? The alternative is to live wholly, eating fresher, purer food, and working out of doors in a way that is freer from stress and rigid routine, and at the same time unarguably productive. And since this life style is almost impossible within a city, to seek it in the country where it is, surely adds up to a valid motive for change.

Around the same time you may detect how helpless you have become in a life style which surrounds you with so many machines, gadgets and services that you begin to feel 'plugged in' to them - almost as dependent as a sufferer on a kidney machine. Less of them around you would free you from worrying about what happens when they go wrong - as they constantly do. Taking control of your own life-support system fires your imagination. You may notice how little control you still have over your own destiny. For you depend not only on unreliable hardware, but on equally fickle people who have become more and more remote in larger and larger institutions.

Shirley and I reacted against this sharply. It seemed to us that the main thrust of city life was *towards* increasing dependence on unreliable people, institutions and technology, *away* from dependence on the natural environment beyond its boundaries. So that we saw the future as a choice between dependence on fickle people or on immutable nature. We decided that we would welcome dependence on nature as eagerly as we would fear over-dependence on such people. A system so riddled with imperfections as Western industrial society did not strike us as worth maintaining - always assuming that it could be sustained. By moving out of it we reckoned we would be contributing less to maintaining it. We knew perfectly well that our actions alone would make no difference to world problems; we simply hoped to set an example. By the same token, we felt that if we were to grow more of our own food and eat it in preference to food imported from distant countries where people were hungry, we would be similarly more comfortable. These thoughts endowed 'comfort' with a brand new meaning for us.

We found ourselves concerned at the way intelligent young people could hardly wait to move the other way – from the country to the city. And we found cold comfort in the weekenders, retired couples and long-distance commuters who were taking their place. We believed that people should try to live as close as possible to their work and where their food grows. And that they should stay more in one place, as they did when towns were small and mobility was a luxury, finding their satisfaction in their work and the company of people around them, rather than in acquiring more and more possessions, in need of constant electronic stimulation and restless mobility. If instead people, alone or in groups, preferably with useful experience and always with careful thought, were to live more simply and with greater self-reliance, and demonstrably more happily, others might follow. Then the gulf between the cities and the countryside would narrow and their interdependence would be understood. Workable alternatives to conventional thought would be recognised. And we might see in our time urgently needed reforms to the present laws on land tenure, so that those with too much land would share with those who have none. If this happened, the land would produce more, and the emptying countryside would be revived. Those who came to it would no longer have to bolster a system which runs counter to their consciences, in order to acquire the cash to buy their way out of it.

You may share our concern or you may not. You may feel an urge to reclaim – as we have – a few of the many thousands of acres of marginal or neglected land which could be supporting families and growing food; or feel angry at the swallowing up of small farms by large ones, when it is an established fact that the larger the farm the less food it grows on every acre. Or you may be motivated more by practical motives than altruistic ones. It matters little. What does matter is your ability to distinguish meaningful motives from ones rooted in fantasy and misconception. Perhaps we can now begin to discern the hollowness in other popular motives for 'dropping out'.

For example, if city or suburban work makes any of us feel ineffectual, how natural it is to play at Walter Mitty and indulge in fantasy, seeing ourselves in the more rugged or glamorous roles of pioneer, farmer, hermit, ascetic, guru. All of which make splendid daydreaming to help hoe row after row of turnips, but are highly suspect as motives for landing you in the turnip field in the first place! And how understandable it is to move because of some inner conflict or a sense of guilt in order to find peace. As Shirley and I have learned, the 'geographical' solution, however, is no solution, for we simply take our problems along with us. Similarly, if our present life seems too complicated, how obvious seems the alternative of returning to the land and an easier, simple life. And how disillusioned we may be! For, as I have already described, any simplicity in living on a little land comes slowly. We leave behind the world of 'instant' this and that to enter a different one of sacrifices, disciplines, patience, and the acquiring of new skills and knowledge. The rewards are abundant, but they are neither automatic nor in any particular hurry to show themselves.

We are likely to be disappointed too if we make the change expecting to find more time for leisure and quiet contemplation. Only those lucky enough to find the perfect place and to enjoy an ample, easily earned income can hope to fall into *that* situation. Most of us must learn to live with an unending queue of farm projects begging to be done; that tranquility and bliss will prove elusive for many years to come. And if our aim is to escape from people altogether, or to win total independence from the System and achieve complete self-sufficiency, we shall be bitterly disappointed. Short of living in a closed community of some size or living as an ascetic, all this is sheer illusion.

Rural self-reliance is possible - for some. It is a goal worth achieving and with some relevance to society as a whole. However, if you set about it with your head full of fantasies and too little understanding, you are unlikely to succeed, and even if you do, then the chances are that it will be the limited kind of success which is measured in mere economic terms; the true rewards of satisfaction and well-being may prove more elusive. As we shall see, this understanding is the key. It embraces knowledge of yourself - your capabilities, limitations and motivations. It includes knowledge of the countryside - its people, farm livestock and other life, its seasons and the vagaries of climate. It implies a commonsense grasp of the realities of money - capital, income and how to acquire the right amount of both. And it assumes that your future will always be in doubt - which is not as fearful as it sounds.

Shirley and I have gained much of this understanding the hard way and the place we farm has rewarded us. But we have few illusions, for we know we walk a tightrope, and both our right to stay here and our capacity to stay here will always be finely balanced. Perhaps we can prevent - in good time - some people from failure; we may even be able to help others to succeed. If so, whether or not our venture survives, it will not have been in vain.

CHAPTER TWO

The Groundwork

Shirley and I have twice left the city for the country. On the first occasion we had no money, but we were young, energetic and impulsive. The war had ended and I discovered that a grateful Government would train me to be a farmer. Shirley had worked on the land during the war and so, when I said I wanted to quit my career as an advertising copywriter, she readily supported me. After two years training on a farm I won a place for a further year at agricultural school. Good jobs were scarce but at the end of the year I landed one managing a pedigree Jersey herd. This I did for two years until we heard of a farm to let. Run-down and sandy it might be, but, longing for a place of our own, we took it and steadily built up our own Jersey herd. Then one momentous day the owner said he would be willing to sell it to us! Nervously we raised the question of price. 'Whatever an agreed valuer decided,' the owner said.

We remember walking over the farm with the valuer. Our arduous work of developing it seemed to escape his notice and he was not impressed with it. His valuation was lower than we had dared to hope for, and so, by borrowing to the hilt, we were able to buy it. We continued improving that farm until we found another nearby with more possibilities. We began negotiations to buy it, and when all seemed settled we let it be known that our first farm was for sale. Now it so happened that our neighbours on either side had long had their eyes on our land, and so we divided it down the middle and - at a handsome price - sold half to each. All was going too smoothly however. Suddenly we encountered a serious legal flaw in the new farm, and lacking the resources for a fight, we had no choice but to pull out. But by now our first farm was sold and we would have another fight on our hands to regain it - even supposing we had the stomach to take on our friends and neighbours. So, with a deep sadness, we sold our herd and looked for another farm which would have all the qualities we desired. We were unsuccessful. As an expedience I took a job as an agricultural journalist - and rapidly found that it was ever so much easier to write about farming than actually do it.

More than twenty years had to pass before we could escape from city life again.

We returned to the land less impetuously than we had first embraced the life, but none the less determined. This time our aims were clearer - and very different. We no longer felt at ease in the industrial society from which we were making a comfortable living. We had learned about ecology. The Third World had entered our vocabulary; its people invaded our consciences; *our* wealth and *their* poverty were not unconnected, it seemed to us. Our own district was growing ugly. Good houses were bulldozed for road widening; a woman was mugged in our street; violence, noise, mistrust were growing - along with the impotence to halt them. The life we led might be comfortable and exciting, but eventually this was not enough to compensate for the ugliness.

Then one day I lost my job - the second time in ten years. Shirley and I discovered something important then: we no longer wished to be dependent on remote people and their institutions. We could not blame them. These people did not know us, and so they could hardly be expected to care for us or even guarantee our basic needs. Instead of hunting for another institution to join, I decided to live directly from my own free-lance writing.

When we returned to the land we did not consider ourselves fugitives. We held a deep conviction that life could be built on different values: concern for the land which supports life; meaningful work which exploits neither people nor nature; more honesty with ourselves and others; less greed and more simplicity.

We returned remembering much that we had learned during our earlier time, but with one essential difference: whereas previously we had farmed with chemicals, this time we determined to grow crops and livestock without them. We had few illusions. We remembered only too well that much of farm work consisted of moving heavy loads from one place to another, and since we intended to keep our use of machinery moderate, we decided to look for a property more like a large garden than a small farm. If we were to eat little or no meat and content ourselves chiefly with food grown on our own land or nearby, meanwhile limiting our other purchases, we ought to be able to live on a very little income. What money we would need must come from my work at the craft of writing.

Our aims seemed commendable, our plans well laid. We had been here before. The passing years presumably had sharpened our judgment. And yet we made mistake after mistake and in consequence we made the going unnecessarily tough. We believe now that we are out of the wood. We have learned from our mistakes and our place is running closely to plan. And yet the future is in doubt. It is probably right that it should be, for uncertainty adds an edge to life, just as complacency dulls it. We are here because we like it, but we are also deeply conscious that we stay because, after all that we have seen and done, there is nowhere else that we feel we could go.

Our mistakes now stand out like haystacks. Our first was elementary: we chose a place which might have been right for someone twenty-five years younger, but not for us. We had planned our move elegantly and the gods appeared to be with us. In the spring of the year, just when we had found a buyer for our apartment, two young friends announced they were returning to Australia and were selling their 15-year-old well-equipped Dormobile van for a song. We bought it. Our intention was to live in it while we roamed the countryside searching for our place. We knew that this was the crucial decision and we had no desire to rush into the first place we encountered. We identified likely districts. They had to be where the price of land was not unduly inflated; they had to be within a few hours of London, so that I might earn some money at my craft; we like hills, so flat landscapes were out; we also like trees; we preferred somewhere fairly warm; and we knew how difficult it can be to make new friends late in life, so we put high on our list those districts where we already knew others of like mind.

The van stood parked in the street outside our apartment, tempting us. Yet we never slept in it. We made a few sorties to see places, staying with friends, but that was all. For one day early on, a friend telephoned from the lovely Wye Valley to say that a house with six-and-a-half acres had become empty. It looked promising, she said, and urged us to come and see it – we could stay the night of course. It seemed to fit all of our preferences; reasonable at the price quoted, within three hours of London, set among hills and trees, blessed with a mild climate and close to friends.

'Of course it all depends on the land,' I said. 'If it has a northerly aspect or poor soil, we must forget about it.'

Shirley agreed: 'And the house. I mean, if it's hopeless – a new box, or some gloomy, draughty Victorian one, or an old one badly modernised – we must be sensible. If only it's old, with a sense of history ...'

'Let's stick to the practical,' I insisted. 'I'm hoping there's a stream near the house, with a good flow and fall, enough to generate electricity from.' I was warming to it, building pictures in my mind. Shirley knew I was, for she was doing the same.

'Don't let's allow ourselves to be too disappointed,' she warned.

Even before we set out in the van we felt confident that the journey would be fruitful. By the time we had left London's dreary suburbs behind we were dangerously predisposed towards buying it.

The local estate agent took us the last few miles in his car, chatting amiably. The place was a bit run down, he explained, but it had attractive features. And possibilities. If we decided we wanted it, we shouldn't take too long to make up our minds, for places like this didn't often come on the market in this district. In fact this one wasn't strictly speaking, 'on the market' yet. So we were lucky to be the first to see it.

We climbed a steep hill and got out of the car. From where the car was parked we could just see the roof of a dwelling below the edge of the road. Beneath us lay a steep valley and beyond it a sweeping view of green pasture and wooded hillsides stretching high into the blue distance. Not a house was in sight. For a moment I was puzzled by a sound like the sea. I looked down past the cottage, past a small patch of garden bright with daffodils, down to the foot of the valley. And there I caught a glitter of water where a swiftly flowing brook tumbled over rocks. Shirley glanced at me excitedly. I smiled at her nervously and we descended the uneven steps that led down to the cottage.

I think it was the trees which made our first impressions so vivid and favourable. Two ash trees grew out of the bank above the cottage, towering over it like the roof of a cathedral. Below a terraced garden at the foot of it were two enormous cherry trees. In the small garden, some ten yards from the front door, stood a single tall cedar. On the banks of the brook, stretching upstream and down, as far as the eye could travel, alders, willows, elders and hazel were coming into leaf. A little upstream from the house, on the far side of a narrow valley, a tall woodland of oak and beech, cherry, sycamore and ash climbed the hill to give way reluctantly to pasture. Higher and higher rose the hill beyond, culminating in a skyline of pines. We turned our eyes to the cottage.

It was small. One end of the slate roof stood level with the hillside from which the cottage seemed to grow. Grey stucco covered the walls. The windows, we noticed, were almost new. Reluctant to enter yet, we walked round to the rear. There a heavy buttress supported the wall - not very effectively, we could see, for great cracks had split its length, and a makeshift wooden wall filled a gaping hole. Stacks of old wood and paraphernalia lay nearby and a lone privy stood guard. More thoughtfully now, we returned to the front.

The agent was trying keys. Presently he opened the door and we followed him through walls three feet thick into a small panelled room with table and chairs, and from there to a slightly larger one, furnished with two easy chairs. At its far end was a diminutive modern fireplace set in wallpaper of simulated stone, stained by damp. I followed Shirley into the kitchen, tiny, dark and smelling of stale food. Up a wooden staircase let into a wall of the living room we found three rooms crammed with furniture. From one of the rooms a passage led outside directly onto the hillside.

'Let's take a look at the land,' I entreated.

Outside again we could see that the cottage was perched almost on the downstream boundary, marked by an ancient, thick stone wall. From the rear of the cottage ran a south-facing hillside, the land flattening at the narrow valley floor. Here, adjoining the brook, pasture grew, but once the hill began, grass gave way to thick clumps of brambles a good six feet high and almost touching each other. From this undergrowth, which nearly covered the hillside, the tops of aged apple trees added to the air of neglect.

Some 200 yards along the track which led from the house and parallel with the brook, a huge outcrop of rough stones projected from the hillside, its surface covered with rubbish and stunted bushes. The agent explained: the previous owner had encouraged the Council to tip rocks and soil from roadmaking work so that a plateau would be formed on top where he planned to build a bungalow. When the rubble-tipping was done he was denied permission to build, so he turned to modernising the cottage instead. Now the outcrop stood, a gigantic folly which almost bisected the property, liberally covered with scrap iron, tyres, rusting household appliances and old bedsteads - all contributed no doubt by an enthusiastic community.

By now the property had mapped itself for us: a curious finger of land some 400 yards long and a mere 80 yards wide, bounded on the

1. Field Gate: *then*

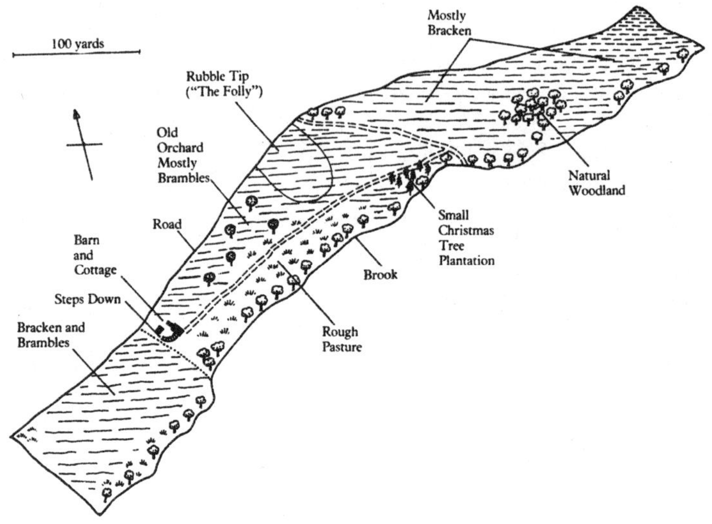

2. Field Gate: *now*

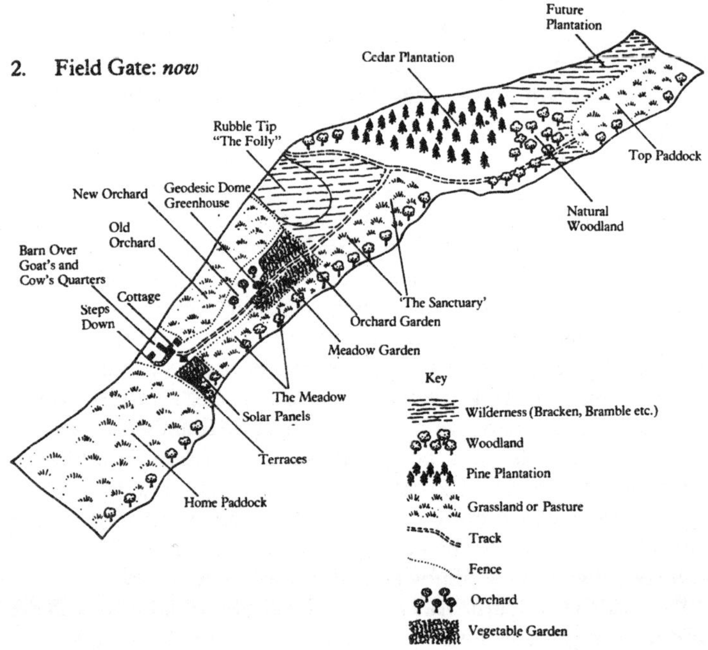

south by the brook, on the north by the road and a side-track; except for pockets of flatter land bordering the brook, all of it was steep; most of it was covered with bracken, brambles, trees and rubbish; only a couple of acres were pasture, and that was clearly rough stuff.

The whole was surrounded by fantastic scenery, and the view from the front of the cottage seized our imaginations; even in the short time that we had been there the light on the creases of the hills had changed many times; it was wild and mysterious, the sort of view you would never grow tired of watching. I summed up the place: beautiful, but far from ideal.

For the next few days we talked of nothing else. We went over its good points: it was within our means; it was handy for London and near friends; it faced south, the cottage was old - yes indeed! - the surroundings exceeded our dreams; I had tried the soil and it was friable and deep; I should be able to generate electricity from the brook. But it was hopelessly steep, covered with too much undergrowth and rubbish; the outcrop was a formidable eyesore; it boasted not one worthwhile gate or fence; and the only track was soft with mud and ended half way. There were no barns or outbuildings to speak of; the cottage itself was an enigma; from the old cider press nearby it had clearly begun life as a barn, in all likelihood some 300 years ago. Since then there had been times when people had cared for it, as the substantial terraced vegetable garden testified. More recently however it had been mutilated in an attempt at modernisation, the project mercifully cut short, though not before creating an overall effect of unrelieved squalor.

We drew up lists headed 'for' and 'against'. We did sums. We drew maps. We had quiet discussions and heated arguments. We went and saw the place again. In the end it came to this: we had fallen in love; through all the maltreatment and neglect, the dormant beauty of the valley and its cottage seemed to speak to us, asking to be restored. It represented a challenge: to undertake such a task and to grow food where none had grown for years, each plot set among trees. They would be beautiful places for work. We would have created something worthwhile.

And so we bought our place.

Your income plan

There is a rough and ready scale of self-reliance: if you see yourself at one end you aim for complete self-sufficiency, growing not only all your own food, but building your own house, making your own clothes, your own soap – the lot; if you see yourself at the other end you have a job and a vegetable patch, and simply grow as much food as you have time to spare. Each of us will place ourselves somewhere between the two, depending on a host of considerations: age, money, aims, aptitude and so on, not least of all how you intend to make ends meet – that is to say, your *economic plan*. You can write this plan down if you like, or simply carry it in your head. It matters little so long as you have given it thought and worked one out, preferably well before you make the move to the country. If you have not, you are taking a grave risk of failure, as we shall presently see.

When the time comes to decide on your place you will have a choice of two basic models: small farm or over-large garden. With luck either can give you what you desire, provided that your aims are clear and sensible, and that you have the capacity to enjoy the hard work, the planning, organisation and determination needed to carry out your aims. Neither model on its own will make you materially rich, nor even self-sufficient in food, let alone other essentials, unless you are fanatic and ascetic. Either model can be made to *pay* for all your food and perhaps for some of your other essentials, but you will need some other income, however simply you elect to live. Just how luxuriously you are able to live will depend chiefly on what you are able to earn, by practising a craft or undertaking other work, either at home or by going out to earn it. There is of course another source of money: unearned income, from investments, property, a personal allowance or welfare. If you are in this class you will fare better, insulated from nature's cruellest excesses and blessed with more time for farm work; but you will not be granted quite the same satisfactions, for in your more thoughtful moments you will know that so long as you suckle the teat of the great breast of that society from which you claim to have severed yourself, you are neither self-sufficient nor fully self-reliant, but nourishing yourself at the expense of others, past or present, near or far away.

You will also be denied full enjoyment of your new way of life if you find yourself spending too much time earning. Except in very bad weather I almost always prefer farm work to 'head work'. For us there is no work more satisfying than growing our own food and knowing that, short of mishaps, there will be some for others.

A source of income, then, is the king pin of your economic plan. Even before you pick your preferred district, consider how well it can be expected to provide you with an income from whatever skills you possess - whether working at a craft at home or going out to work, seasonally or part time. You will probably require about half the income you earned in the city, but even this may not be easy - as we shall see in a later chapter.

Unless you do your homework carefully you may find yourself trapped, for the more time you spend earning money, the less you have for developing your place and for routine growing - in other words, the less self-sufficient you become. Conversely, if you buy a place that is too run-down or difficult to work, the less time you have for earning, as we have found to our cost! Either way you walk a tightrope.

You can buy time by employing labour and machinery

You have to accept that all along you operate within three principal constraints: time, energy and money, and they are interchangeable. If you have chosen a district where you can earn enough money, you can buy time, and you can do so by employing others' labour and machinery - unless of course you feel that in doing so you are exploiting them. Alternatively, if you have abounding physical energy you can keep the momentum going by undertaking more on your own land and so save the expense.

How much land?

People keep asking us the same question: 'How much land do I need to become self-sufficient?' And every time we have to give the same answer: 'It all depends.' You see, the acreage varies according to the quality of the land, the climate of the district, how much meat and dairy products you will eat, how many people in your unit, and how self-sufficient you intend to be. As a rough and ready guide, each person in Britain ought to be able to get all his staple food requirements from half an acre of really fertile land, in a district where wheat will grow and the growing season is long. However, if his land is fair to average quality and the climate possibly less benign, he will need an acre. If it is marginal land he will require a couple of acres, and still not be able to grow all he would like - cereals especially. All these figures assume he *can farm.*

It would be helpful if there were figures showing costs, land requirements and monetary returns for each crop and type of livestock. There are not. Ask a dozen 'experts' to calculate them and you will get a dozen different answers. They vary from farm to farm, district to district. Best plan? Ask your neighbours.

Now few of us are able to be completely self-supporting, and as I keep saying, there is little sense in trying to be anyway. This means buying in food for people to eat and almost certainly for livestock too. Once you do start doing this you can manage on less and less land, so that at the 'easy' end of the self-sufficiency spectrum you will find that a plot of good land, well looked after, and a mere 300 square yards, will probably supply an average family with all its vegetable requirements except for potatoes - and it should provide at least some of these.

The best you can do is to form some idea in your head of the size of holding you can visualise yourself managing well and happily. Do you see yourself on a small farm of perhaps 15 to 50 acres, or on a large garden of, say, one to three acres - or something in between? Before you can decide, you must have a clear idea of what you are doing and why; for if you do not, you will probably buy the wrong place, the wrong machinery, the wrong stock. You will end up trying to grow square pegs in round holes.

The over-riding constraint within which you operate will be the size and productive potential of your land. Robin Clarke, writer and self-supporter who farms 23 acres in Shropshire, comes down heavily in favour of a small farm rather than large garden. '... You will be stuck with whatever land you buy,' he emphasises. 'Nothing will make it any bigger and the chances of buying more nearby at a later time are slim indeed.'

I do not disagree with him, but I would say: if you want to farm, by all means buy enough land to do so, but first satisfy yourself that farming is your aim. It is not for everybody. The only way you can tell is by working on farms over a period, not just for odd weekends in the summer, but all the year round, especially in winter when rain stings your face, cold numbs your fingers and mud makes the going hard. Gardening demands you to be out in all weathers too, but the more livestock you carry the longer you must take over the daily tasks of husbanding them whatever the weather.

If you see yourself as a farmer and you can both buy the necessary land and leave yourself enough working capital to spare, you will enjoy the beef, pork, bacon, lamb, dairy produce and cereals it will yield. But if your interests veer more towards growing on a small scale, with perhaps more time to spare for a satisfying and profitable craft, you might still opt for less land. What matters most is your over-riding interest - farmer or gardener. If you feel overwhelmed by too many acres you may become frantic or bewildered; but if you have farming in your blood and you are forced to dig and hoe for the lack of land to run a few cows or grow a crop of wheat, you will be a sorry spectacle.

You must fit your land and your land must fit you, for the two of you will build a partnership as close as any marriage. You will change the ecology of your land; but your land will determine the tenor of your life. What you grow on a small acreage will generally be less than on more acres; therefore, as a self-supporter, you may have to be content consuming less meat or none at all, and possibly less dairy produce. In fact unless you have a handsomely paying craft, you may find yourself consuming less of everything. Since that may not please you, consider the potential of the land, weigh up your capital and earning prospects, examine your bent - farming or gardening and decide how much of an ascetic you are at heart. If in doubt, do not stint yourself on land.

Our own plan

With each new season more of what Shirley and I are doing fits into place. The other part of our economic plan - what we do on the land rather than how else we make ends meet - becomes clearer. Perhaps our plan is best illustrated by how we do our accounting. We divide expenses and income into two categories: food and 'other'. We ask the land that it should grow most of our own food and yield a surplus for others - if it did not we should not feel comfortable about accepting from others the rest of our food, and the artifacts and services by which we also live. This surplus we sell or barter, using the money so gained to pay for what it costs to grow the food: fuel, seed, maintenance of machinery and equipment, fertilizers, feedstuffs, together with a share of running the van, since it is used for food production. We also aim for the surplus to be big enough to pay for any food we buy. Everything else has to be paid for from the income I receive from writing: rates, stamps, telephone, building maintenance, land development, pocket money and so on.

It took time for the land to start contributing its full share, and not surprisingly. Remember, when we came, its steep hillsides were growing virtually nothing. With undeserved and unexpected help from people, young and less young, we hacked at the wilderness and rough pasture to create an acre of fruit and vegetable garden; we cleared two and a half acres of bracken, brambles and undergrowth to sow new, improved pasture; we planted 1,000 red cedars on land too steep for much else; we erected a geodesic dome greenhouse; made tracks, built fencing and put up gates - all where there had been none; we made storage for hay, tools and machinery, and quarters for livestock; we spread - mostly by hand - tons of lime and rock phosphate to correct revealed soil deficiences; we stocked up with Anglo-Nubian goats, a Jersey calf and Arbor Acre hens - all chosen for their docility among other points; and we plumbed in three square metres of solar panels to heat water for house and dairy.

All this happened in the short time of two years. Our diary reads like a cine film out of control! By then our place was already growing most of our own food and a worthwhile surplus for others, despite two dry years - one of them the worst on record. We can think of few satisfactions greater than that of growing food on land which was recently derelict. Every self-supporter must ask himself the question: 'Is this land growing now more than it did?' and if it is not, he must ask why.

Our satisfaction is the greater because our place is even more beautiful now than it was before we came to it. We have to check a sense of pride, reminding ourselves how much we owe to others in the doing of it. When I show each of our visitors round and point out the many small miracles which have happened here, I stress that I do so not out of self-congratulation, but simply to explain how it was that so much happened in so short a time. Similarly, Shirley and I, though we have bought this land, do not feel that we own it; we feel we are merely stewards of it, to protect it from ill-doers and guide it towards an abundance which, we hope, will distinguish it from less fortunate land long after we are forgotten.

Your plan for your place will probably be different from ours. Any plan, however, must be intimately related to the way you intend to live and hence the principles by which you believe others should try to live. It is something to decide in outline even before you settle on your land, as indeed we did, for we understood clearly, long before we found our place, the only life style that would allow us to be comfortable within ourselves.

We erred in our plan, not because we bought too big a place, but because we failed to take into account either the cost and effort of developing it or the difficulties that its steepness would cause, once the development programme lay behind us and we concentrated on farming it. Perhaps our ingenuity - a faculty now developed beyond all expectations - will come to our rescue. If it does, splendid; if not, we may remain materially poor and physically spent, but we shall have our views, the enchantment of our valley and the satisfaction of achievement. Shirley has been aware of our blessings all along, I am slower in learning to count them.

How far to go?

Pick up a book on self-sufficiency or survival and the chances are that it will cover not only farming and gardening but killing and dressing meat, cheese and butter making, wine and beer making, bread-baking, building, energy from sun, wind and water, beekeeping, cooking, spinning, weaving and dyeing, pottery, carpentry, tanning, soap-making and blacksmithing. One I know tries to tell you how to make bows and arrows, musical instruments, and paper and boats as well! Many people can be forgiven for believing that the aim of the game is to take on the lot. It certainly is not. No one I know has, and anyone outside of a community who tries to do so will be courting disaster. So just how far should you go?

We find that converting a house, developing a property, growing vegetables and fruit and keeping two or three dairy animals - on a scale just big enough to feed us and produce a small surplus - as well as earning a small income by writing, have all added up to more than a full time job. One day perhaps we shall run out of ideas on improving the property; one day we might have the cottage looking immaculate. Then I suppose we shall have the time to weave, make soap, make clothes and carry on other crafts which would help to make us more independent from the System. That day is not yet with us.

The all important thing is not to attempt too much at once. Plant too much too early, for example, and you are sure to find that you will not be able to hoe it, net it, dust it, water it, or harvest it properly, and you will finish with meagre crops - probably no more than if you had planted less and looked after them better. What is more you would have incurred less work.

If your house and buildings are right for small scale mixed farming or vegetable growing, if your gates and fences are sound, your land in good heart and everything is growing nicely, you should have time enough to attempt one or two crafts in addition to what you need to do for income. What you tackle first should surely depend on what you need, what comes easily to you and what you feel like doing - probably in that order of priority.

If part of your credo is a rejection of contemporary industrial society, you are bound to exercise caution in accepting and using its artifacts. After all, you are on dangerously thin ice criticising it with one breath - because it exploits nature or people, because of the trend to nuclear power, because goods are shoddy or expensive or, because they make you too dependent - and then with the next breath phoning your neighbourhood agricultural agent for the price of a tractor or rotovator, or ordering *any* mass-produced goods at all. The uncomfortable truth is that the more you use, the more you must temper your criticism. One way out of course is to make your own. It sounds an attractive solution, but again reality can prove to be uncomfortable. Home made things are not *easy*. They take up a lot of time. You have to hunt out or prepare your raw materials and make or buy tools and equipment. You will be slow while learning and make mistakes and rejects. Even when proficient you may find your chosen craft too time consuming, possibly tedious. You may finish up 'robbing Peter to pay Paul': there you are beavering away potting or basket-making, only to find you have so little time left to grind flour and bake bread, that you have to take out the vehicle and buy a wrapped, sliced, white, plastic loaf; or you opt for candle-making, only to find that your free time is in the evening, and you are burning up more energy from the mains than you will generate. All simply to make a point!

The real point is that some things are best made by experts or with machinery and - I freely concede - even from plastic rather than natural materials. Since we cannot make all we need, we must discriminate. We should concentrate on our greatest needs and the things we enjoy doing, because we turn out a better looking, better working article than some machine, and hopefully with fewer undesirable side-effects.

In a big enough intentional community, well blessed with land and raw materials, preferably with steady wind or water power, you may find that good organisation grants you the time you need in order to come close to complete self-sufficiency. Certainly the spread of talents which should be available would prove a wonderful asset. But the smaller the group of people, the harder the ideal is to achieve. In any case, what is the point? Surely it is enough to reduce your excessive dependence on others, on machines and cities? You have no need to eliminate dependence altogether. If you do you will simply see fewer and fewer people. And if this is how you believe life should be lived you are in effect proposing that each family, each community, village or town should build a wall around itself. I would argue instead that a sensible degree of mutual dependence makes life easier and is more fun. I see no point in trying to put the clock back. I see great point, however, in curbing our excesses, in freeing ourselves from over-dependence on remote people, uncaring institutions and on machines. Ghandi declared that so long as a machine was useful it should be used, but once it became *necessary* we should manage without it, for if we did not it would enslave us. Ghandi also welcomed trading between villages. What he saw as unjust and dangerous, in contrast, was the inequality and lack of understanding between countryside and city.

Total self-sufficiency may have some academic value, it may fulfil a psychological need in some of us; but as a social experiment and a way of life it has little relevance to today's pressing problems. It is certain to be such long and arduous work that those who seek it will be so concerned with sheer survival that they will have no time for much else. And that is not much of a life.

Personal qualities

By now you will have gathered that rural self-reliance is not everybody's cup of tea. You need to be an unusual sort of person to enjoy it and succeed. So what are some of the personal attributes that are likely to stand you in good stead?

Obviously you must be physically in good shape. Less obviously perhaps, you will find that some brains come in handy. It takes some intelligence and plenty of imagination to see a place as it should be when it is not. Chances are that what you buy will be run down, or some piece carved off a bigger farm, or something the locals do not want - and for good reason no doubt; or a specialist place - chicken, dairy cows, pigs or some other mono-culture ... whatever it is, it is unlikely to have been equipped for self-reliance. So it will probably have unsuitable buildings, fences and gates in the wrong place, grass where you will want vegetables, and weeds where you will want an orchard. The house will have none of the shelves, cupboards and working areas you will need for all the food you intend to preserve and the good things you will want to bake. It will be too flashy here and too primitive there. You will need 'five-year vision', that special kind of eyesight which can catch a vision of how things *should* be, and hold the vision - though not inflexibly - for the long years it will take to turn it into reality.

For that to happen you will need energy for work, both the kind you do with your head and the kind which uses the rest of you. And since all will not always go as you have planned, you will also need plenty of determination, resilience and resourcefulness.

Unless you are something of a stoic, you will soon go under. Maybe in your very first year you will lose your hay crop from incessant rain -as we have, or your potatoes to 'blight' or 'blackleg' - as we have too. If you cannot take such setbacks nor live in the shadow of their possibility, better to stay in a steady job and the deceptive safety of the city, for there you can more easily escape from trouble. On your farm, with livestock and growing crops, you can get away only if some responsible person takes your place - and that is not as easy as it sounds.

We all procrastinate but if you over-indulge, you will store up trouble for yourself; we all panic at times, but to live in a chronic dither is to get your priorities wrong. You will need help from other people all along, so a generous sprinkling of friendliness in your nature will not go amiss. Nor will a touch of tolerance, for you will often be misunderstood, and while you need not seek the approval of others as avidly as you may have done in your old life, you will not want to incur disapproval unnecessarily. Yes, you will certainly need help, and your best source will be from your partner - or partners. If one of you has nagging doubts and is coming along out of a sense of loyalty or in a spirit of 'let him get it out of his system' then not only the relationship but the whole venture will founder. What you are doing will strain the bond that binds you, so rigorously that if it has any serious flaw it will surely break. On the other hand, if the relationship is sound, you will draw more closely together than perhaps ever before, children and all - one of the greatest joys that this way of life has to offer.

That you are choosing to live and work in the country rather than the city implies some interest in nature - hopefully one based on partnership rather than exploitation. If you think your role is to 'tame nature', better think again: you will make more progress trying to understand her laws; to appreciate her in all her forms - from bacteria to badgers, in sunshine and slush. I am not suggesting that you can set up an absorbing concern for nature in advance as a precondition for moving to the country - this is a harvest that ripens slowly. However, if you can cultivate a humbleness in your own nature: to be content on your own; to be at one with yourself rather than constantly craving outside stimulation, you will be greatly helped.

Such quiet study is paramount, but clearly it is not all. There is work to be done - which is where your physical strength and fitness come into their own. And these must be backed by some practical skills, not only in farming and gardening, growing crops and husbanding livestock, but in maintaining and servicing machinery, and in the practice of some money-earning craft. And so, if you do not have any such skills before you decide to make the move, you would do well to take steps to acquire them - possibly by going to day or evening classes and by working for a while on farms, so as to test whether you enjoy the practice of them and can do so with some aptitude.

Now most of us have enough of the qualities for making a success of this way of living; the trouble is that our high-rise, high-speed, city-centred society has just about killed them. They have to be rediscovered and nurtured to bloom again, and this is an exciting project to undertake as you explore the possibilities of living differently.

So far I have concentrated on skills and traits, but there is at least one essential asset which falls into a different category and that is money. Yes, money. For as matters are, you will almost certainly have to buy not only the property itself but things you have not previously considered, and in quantity, firstly to get started and later to run your place properly and without undue anxiety - as we shall discuss in a later chapter.

Money you will need; strength, brains, skills and sensitivity too. But unless you have one other attribute you will hoe a hard row. When the house cow has put her foot in the milk bucket; when the vehicle breaks down in a rebellious heap of metric nuts and you have nothing but Whitworth spanners; when you collapse after ten hours haymaking only to be invaded by your recent city neighbours with four fractious, hungry children ... on such occasions you will truly need your sense of humour. If all you have is a weakly seedling of a thing, nurture it - perhaps it will flourish in your new life; but keep it alive at all costs, for one day it will see you through at least one crisis.

One self-supporter, who seems to agree with me on most of the personal attributes you need for success, is Sedley Sweeney, whose 10 acres in West Wales keep him and his wife in most of their needs - and a few luxuries. He stresses the value of incentive too, which he says '... grows from a realisation of the true state of the world. We must learn to survive or go under. It may not be long before the 'soft option' of the welfare state no longer exists to come to the rescue of those who fail. To avoid failure he believes that every aspect of your project calls for discipline: 'To acquire the habit of strictly frugality before it is forced on us; to keep really fit for the long hours and heavy work; to keep mentally fit to hold onto our aims and priorities when soft options beckon; and to acquire the skills necessary for every facet of the work.'

Unpopular though the word is today, I would certainly agree with Sedley Sweeney about discipline. To be a self-supporter calls for a different kind from the *imposed* discipline you encounter in a nine-to-five job. It is all up to you. A friend phones you in the morning to come over for a chat and you would like to, but it is just the time for planting onions, before the weather breaks; or you've caught 'the bug' that's going around, you feel like death but the wood pile is exhausted and unless you saw and cart some, bread cannot be baked nor the house kept warm. It is a *self-discipline* that you have to foster.

To anyone contemplating this life I would say: 'Look before you dream, work before you leap.' It is a great life for some. To be sure it is right for you, gain that experience on someone's small farm for at least a year first. After living it for a while you will discover that, though you may be materially poorer, you are spiritually richer. The life you've chosen may be hard, but that is good. Like other species we have evolved in struggle. Three million years or more of struggle warding off predators, coping with the elements, have left us with a deeply felt urge to continue to struggle. It is more deeply ingrained in our natures than the brief years of city comfort that comprise our recent history. And so the paradox is that we still require struggle to feel satisfied. Yet how can we feel satisfied with a life filled with the struggle for trivialities such as status and acquisition? To be fulfilling it must concern something more real and basic. Survival - warmth and food - this is the stuff of self-reliance and it concentrates the mind wonderfully. It releases us from the problem of 'overchoice' - that affliction of affluence which leaves us confused and neurotic, unable to decide which of umpteen options to take up. If survival is at stake there is no problem of agonising about when or where to take a holiday, whether to change jobs, whether or not to redecorate the house and what colour, what to do if we won the pools. When 'overchoice' flies out of the window and you accept that you are staying where you are, with one aim, it is all for real; there is no room for illusion, little time for fantasy.

Breaking rules

It is not a bad idea to strive for perfection - so long as you accept that you are unlikely to find it. To aim for purity in your way of life is a noble ideal, if you can stop tormenting yourself with guilt each time you slip. There is no harm in having a dream about your place: how it will look, how you will run it and what your aims for it will be. No harm at all, so long as you recognise that there will be times when you will be tempted to break rules and take short cuts. For you will yield.

Shirley and I had a dream for our place. We still have, though it is not quite the same one. We are closer now to realising it and we are sustained by seeing the barely possible take shape more quickly than we had dared hope for at the beginning. We have few illusions about the achievement. As we broke rules and took short cuts we reminded ourselves that we were in our fifties and able to count on our fingers the number of really active years left to us. We did not fancy standing in the wings while others played our parts for us: we wanted to participate. And this we could not have hoped for if we had not resorted to bulldozer, heavy tractor, paid labour, contractors, high technology, and - at least once - *chemical fertilizer!*

Imagine the situation. There we were at the beginning with our ruin just habitable, acres of derelict land around us and the growing season barely three months off. The hillside near the cottage was thick with tall, flesh-tearing brambles; our flatter land below, where we might grow vegetables and fruit, was a mixture of rocks, rubbish, undergrowth and old, weedy pasture; beyond this a half-acre of gently sloping, south-facing land, was covered in unkempt, mis-shapen Christmas trees; and at the top of the property beyond the steep, bracken covered hillside was another three-quarters of an acre of sunny, flat land – with one drawback: you could not reach it except through an assault course of rocks, undergrowth and tumbling water.

We were possessed of a vision of our neglected place, made beautiful and productive: small plots of intensively cultivated land, abundant with colourful vegetables, crops and fruit, set amid dappled green wilderness and plantations of sturdy young trees; a firm track winding through this intimate landscape, upstream to the far end where the steep, wild hillside would suddenly give way to a vista of lush pasture set aside for hay; the whole bounded on the high side by a road – largely hidden from below – and on the low side by our brook, bordered by ferns and wild flowers, splashing cheerfully over clean stones under shady trees.

Between the uncared-for present and the bountiful future lay one obstacle – work! Land to clear, rubbish and rocks to shift, soil to cultivate, crops and pasture to sow, trees to plant, a track to build, fences and gates to make. We looked at our immediate resources: a five-horsepower rotovator and trailer, a chain saw, a basic array of spades, forks, other tools – and ourselves. By the glow of our fire in the evenings and as we tramped over our land by day, we talked, surveyed and planned, doing sums, then doing them again, always examining our motives. Our broad plan seemed to emerge clearly enough: we aimed to grow as much of our own food as possible and a surplus besides; this we would do on all the land flat enough to cultivate and sunny enough to grow crops; the rest would be divided among fruit trees, timber plantations, pasture and wilderness; we would run hens, goats and possibly a house cow. We would use as little machinery as possible and apply no poisons to the land. There would be no 'bad vibes' on our place. Shirley and I lived together in harmony; we wanted the same harmony to pervade our place, and for people to sense it when they came. We hoped that this land would be an example to others of another way to live.

Our lives had encountered an unexpected twist too. Gone were any dreams we might have entertained of instant self-sufficiency and time for leisure and contemplation. Somehow or other we had landed ourselves with an experiment in reclaiming derelict land – the kind of which there are countless thousands of acres throughout Britain and beyond, the kind which you will find in corners of larger farms and vast estates, the kind which agribusiness has no time for. And we felt obliged to demonstrate the results of our experiment if it was a success.

The prospect was exciting and not a little daunting.

For one thing, both overshadowing us and stimulating us in our endeavours was our deeply held conviction that a world food shortage was imminent. Everything pointed that way: exploding population, 'green revolution' failure, continuing soil erosion on existing farmland, no more easily cleared large tracts of arable land, a fuel and fertilizer shortage, and all the risks of crop failure inherent in monoculture. Added to which, Britain faced bleaker prospects of buying food from abroad, the scarcer that food became.

We change our plan

At some time the concept of a 'five-year plan' entered our discussions and took hold. Why we picked on five years is not now very clear. Perhaps it simply sounded rather grand. Alternatively it may have had something to do with the threat of food shortage. Five years probably seemed a likely time for the screw to turn; it also promised to be the earliest date by which our programme could be complete. However, no sooner had we grown used to this fairly leisurely programme - hard though it might be - than we woke up to the new reality of inflation. Almost overnight our well laid plans crumbled. This was a factor we had failed to reckon with. For each year - indeed *every day* - our precious cash reserves were evaporating; each year they would buy less. We could neither cushion ourselves by living off them while we threw ourselves into our land development programme, nor count on their buying us what we needed to see it through; not unless we bought what we would need virtually right away while our money still possessed some real value.

'We must telescope our five-year plan,' I declared grandiosely. Shirley put on a worried look: 'How?' she asked significantly. 'I was afraid you'd ask that,' I said, playing for time. We fell silent until we felt ready to carry on discussing the prospect, and found happily that we agreed on the change of plan - as usual it had already been in Shirley's mind. Slowly, a way of telescoping it evolved.

'We shall need help,' she put in, 'Even if it costs money.'

'Even if it means heavier machinery than the rotovator!' I reflected sadly. 'Even if it means we're less self-reliant for a time.'

Well within three years our five-year plan was accomplished and at last we hoped that our frenzied activity could slow down. We were poorer financially, but more secure by having transferred our reserves from bank to land. Not that the transfer made economic sense in the short term - nor perhaps in the long term either. That depended on whether or not our fears of inflation and food shortage materialised. But the basic work was done, and we could look forward to continuing it actively ourselves - instead of looking on one day from our rocking chairs!

There was a price tag attached to our accomplishment of course. We had compromised some of our ideals and our consciences had been tweaked. We had winced as the bulldozer we hired crunched remorselessly through our woodland to make the track, even though we had a tree-loving driver who picked his route carefully. We had worried when a neighbour lumbered in with his giant tractor to start ploughing our top field - the proposed hay meadow. And we weakened when later an even more massive bulldozer nudged stones

Clearing boulders from land destined for pasture

of unexpected proportions out of the craters it scooped, and deposited them by the hundred along the bank of the brook. Yet all this had the drama of power and immediacy. Later we suffered nagging doubts as one helper after another - paid and unpaid - came and went, over the months, each progressing the programme almost imperceptibly towards its completion. Perhaps the greatest dilemma centred round the costly herbal ley, and especially its clover content, on the Home Paddock in front of our cottage.

This one-and-half acre field was originally a jungle of tall bracken and brambles. We hacked at it, mowed it with an Allenscythe, raked it, burned the brambles and composted the bracken. Now cleared, we could see this would be productive land, a lightish red loam, deep, so deep, and mercifully free from stones. Yet it was some of our steepest. Here Owen, our good neighbour, ploughed. He would drive his great tractor along the top where there was a flat, narrow headland, then turn sharply, and as the machine lurched forward drop his heavy four-furrow plough. At this critical point the plough acted as brake,

Ploughing a 1-in-3 slope after hacking 'jungle'

biting into the soil, tearing through the mass of bracken roots to slow down the tractor on its downhill journey. Sometimes even so, it gathered speed ominously so that Owen was forced to brake, and then the wheels would lock and he would continue ploughing, careering to the bottom, drawn by the force of gravity. The furrows complete, he would slow down, turn and raise the plough with the hydraulics. Then he would climb diagonally up the slope to the headland for the next run, the tractor leaning over precariously, now seeming to *defy* gravity rather than make use of it. Shirley had to look the other way. I was torn between unqualified admiration and the need to get on with other pressing work.

In time the field was ploughed and it seeemed that the danger had passed. But then came the time to harrow it. Without the breaking effect of the biting plough the tractor hurtled downhill alarmingly; the only saving grace proved to be its speed, for the job was done in a tenth of the time it took to plough - and Shirley was able to function again.

All this happened in November. We planned to sow the field in the spring - by hand, on the contour with an ancient borrowed 'fiddle' and to harrow it lightly by dragging brush-wood tied to a crosspiece in the old style - again on the contour. And the pasture *had to take.* In all conscience we could not ask our friend to risk life and limb in ploughing a second time if the seeding failed to germinate or died away afterwards. And with bagged nitrogen soaring in price, and 'ethically unacceptable' as an annual event anyway, above all we had to ensure that the clover in the mixture would live and thrive. Now there is one element clover needs above else and that is phosphorous. Without it, young seedlings will not grow. Because soil tests had shown that our land lacked phosphorous, we had arranged for our friend to spread rock phosphate - natural rock, mined and finely ground - and this would take care of the deficiency in the long term. But rock phosphate had one major disadvantage for this event: available phosphorous was released very slowly, but young clover-needed it right away. There was only one way we could see round the difficulty: we would get our friend to spread superphosphate as well as rock phosphate. Superphosphate is water soluble and gives a quick boost to needy seedlings, but as an unnatural, man-made chemical it is unacceptable to those who stick to organic methods. So what? we asked ourselves. That we should be so pure! Had we not already sinned by using heavy machinery and letting a paid helper take heavy risks? Moreover, we were not intending to force our land by repeated application, wherein lies the chemical's chief harm. This was to be a once-only event. And so the superphosphate went out, and despite a spring and summer of unprecedented drought, the clover lived.

Since then we have modified our views on purity, tempering them with expedience. In short, it seems that the chief danger of using technology, and the chemicals and machinery which are its fruits, lies in lack of discrimination. To use a potentially harmful method when a benign one will do is surely an act of folly. So too is to become 'hooked' on applications of a chemical year after year. For special circumstances we feel that at least *some* rules are made to be broken - whether to germinate one acre of precious clover or to grow food on several acres to bring land into production two years early in anticipation of a severe shortage. If we have erred, may we be forgiven!

CHAPTER THREE

The Move

You open your eyes sleepily to a strange light. Gentle, new sounds float in through the bright window, and the cool air wafting on your cheek has a different smell. The unaccustomed room is barer than you expect. With the shock of wakefulness you remember: you have arrived at your place, and the significance of this day surges into your consciousness. The events of the move tumble in one after another; you crane your neck to look round the room, and the boxes and clothes dumped on the floorboards prove that it really happened. You savour the delicious moments of anticipation before rising, until a sudden sense of awe overwhelms you: from now on everything depends on you: nothing will happen unless you make it happen; beyond that window lies land, latent with fruitfulness and you must waken it; strange objects are strewn around this house and you must master them; the freedom which you longed for has arrived, and with it new responsibilities. You catch your breath with the wonder of it all, and your heart sings.

Let no one underestimate the cultural shock of leaving the city to live on a little land. All of us have spent some time in the country - on holiday, to work a while on farms or perhaps as a commuter - but such experiences are small preparation for your new responsibilities. So you would do well to equip yourself beforehand, with skills and tools and clothes and attitudes of mind appropriate to the task ahead.

Shirley and I had two years' warning of impending change to our lives. My boss had told me well in advance that I could expect to be made redundant; and even though we had not decided what to do with the rest of our lives when the blow fell, we knew that it would mean living on less, and so we began to live accordingly. In my last year as an employee we saved one third of my take-home pay. Food was our chief target. We ate less and less meat, rejecting the most suspicious animals first. Veal was already taboo, broiler chicken went soon after, for we knew the cruelty of factory farming. Beef and pork soon followed them: somehow to eat an animal fattened on grain or fish meal imported from countries where people were hungry took the edge off the flavour. We stopped buying expensive imported fruit and vegetables and anything out of season. We had never been ones for convenience foods, but we now forbade them, cans, plastic packs and all. Cream cakes and fancy biscuits we reserved for celebrations. We shopped around; we bought in bulk; we ate less, and when we entertained - now more rarely - we served two courses only. Progressively, we cut down our moderate alcohol intake until it was wine only, for celebrations.

We sold our car and stopping going out for the sake of it. No more films or theatres. Until we became used to our new inputs we relied on our ancient TV for entertainment. We stopped buying *anything* that might be irrelevant to our alternative lifestyle - whatever that might turn out to be. When we did buy, we bought as much as possible secondhand. I grew a beard, and that saved razor blades, shaving cream - and precious time. For there was much to do. We had to get fit, physically and mentally, bracing ourselves for the challenge ahead, as yet unclear, but assuredly unlike the soft, parasitic life style we had been leading for the past ten years or so.

After the blow fell, and I had resolved never to become so nakedly at the mercy of remote people again, a whole year passed before the idea of living on a little land crystallised. From then our training intensified. We had a lot to learn. Having farmed before we had less need than other would-be self-supporters to taste farm work; if not we would have probably joined WWOOF - Working Weekends on Organic Farms - and spent weekends and longer periods tasting the life. We did taste life in communes to discover whether we would be more effective living and working with others rather than going it alone. However, after much agonising and a touch of drama we decided we had left our move until too late in life to make the profound changes that community would demand of us.

We read books and magazines avidly and took steps to meet people already living in the alternative way we were planning. We learned the value of testing our ideas and prejudices by discussing them and getting into arguments with these new friends, either in each other's homes or at conferences and happenings. We could have done more to acquire new skills. Shirley brushed up on her bread-making and improved her vegetarian cookery, but I might have learned motor maintenance, building, plumbing, carpentry, welding or any one of a dozen other useful crafts. However, I failed to do so, claiming in my defence that my time was full, relearning my old, money-earning craft of writing.

There was much to buy, too. We guessed that once we were several miles from a town, busy with the job of farming and gardening, shopping could be as difficult as it would be unwelcome. So we took advantage of the city for its specialist stores, cut-price shops and opportunities such as jumble sales to buy tools, equipment, clothing and books. We made savings; we also made mistakes. I learned that old tools were worth hunting for, that cheap tools soon broke; we bought curious things we still have not used - waterproof working clothes which defied us to work in them, a vanity basin, ceramic bathroom tiles, quantities of screws in improbable sizes and old-fashioned flat irons which, although never ironed with - for ironing is a vice Shirley has given up - have nevertheless proved useful for propping doors open. All the same we were glad to have a nucleus of garden and farm tools and work clothes from the word 'go'.

Choosing the district

Looking back we can see that, while we were diligent in preparing ourselves for our new life, we were less than thorough in choosing where to live it. Ideally there are two stages: firstly deciding on the district and secondly picking the place. We relied too much on chance, opting for a district which we already happened to know. With the wisdom of hindsight we can now list many of the considerations that would guide people more sensible than we were.

For each of us, some points matter more than others. Remember that latitude, altitude, rainfall and soil type will all affect what you can grow and sometimes will affect what you can sell. The trouble is that perfect places cost more than the sort of places most would-be self-supporters are likely to be able to afford. You, too, will probably have to make the best of *marginal* land, imperfect in some way that makes it less desirable for adding to its neighbouring farm, or unattractive to the commuters and retired couples who dominate the market for small farms and large gardens. If you have a limited capital, something has to give. You may be led to a district of mountains. Now the side of a mountain will afford you glorious views, but it can have drawbacks – possibly more daunting than restriction on what you can grow. Elizabeth and Alan West went to live 'the simple life', 1,000 feet up a mountain in North Wales, and they soon found that, although from May to October their cottage is dry and sweet smelling, with colder weather the walls of every room stream with condensation. Explains Elizabeth: 'Any foodstuffs, clothing, shoes, etc., left touching an outside wall will become saturated and sprout beautiful little tufts of white fur, and every article in all rooms except the kitchen will be affected by damp. The bedroom over the parlour is uninhabitable in the winter. Condensation drips from the sloping ceiling onto the bed, and our breath condenses in clouds around our heads. So we move to the bedroom over the kitchen. Here it is necessary to have the bed in the middle of the room.' Although their neighbours never admit it, they all have the same problem; no one – the Wests included – has yet found an answer.

One simple rule applies to any district, any property: don't be carried away by what you see on a fine summer day, for you may spend nearly half your life in wintry conditions. Then everything looks and feels very different. If altitude can inhibit you, so can latitude. We have good friends, keen self-supporters in Aberdeenshire, who shrug off winters of biting cold, blizzards, darkness by mid-afternoon and nothing growing for months – all of which would send us scuttling back to the city in no time. Although some crops grow better in northern latitudes, generally speaking the further north you go the fewer crops will grow. However, as our friends pointedly remind us. 'Some people actually choose to live where runner beans won't grow, for financial rather than masochistic reasons. And a diet deficient in runner beans isn't all that much of a problem by comparison to being able to afford six acres and a home and buildings, is it?

Beware excessive rainfall, too. If you pick the wetter parts of Wales or the West Country, for example, you will not grow good wheat except in unusually dry years. Roughly speaking, the more scenic the landscape the more problems: slopes are hard to work and sensitive to erosion; trees attract birds, which are welcome neighbours until they massacre your winter wheat and spring fruit buds; lush growth spells beauty … and often successive deluges. So if beauty inspires you, strike a balance: choose something between closely settled, treeless, drained fens of fantastic fertility and that enchanted deserted valley, which nestles secretly below hanging crags and wild mountains in the land of nowhere. There could be a reason why no one lives there. Perhaps the soil is no good. Remember that wherever you are the quality of the soil is paramount: it must either be fertile or be capable of becoming so. Of course remoteness has its attractions, especially for anyone suffering from an overdose of traffic, neighbours and noise. But you may be a long way from friends and family, and while you will almost certainly find them dropping in at all times, you may not be so eager to down tools and face long trips to see them. Remember too that you will be a long way from markets for your surplus produce or craftwork, a long way from work when you need to earn cash. You could find yourself too far from anyone at all. Then one haymaking time when you have strained your back and you can barely lift a spoon, let alone a scythe, or one autumn when there is just a week to plough and harrow before the land is too wet and cold, and your tractor expires, you may wish heartily that you were close to friendly people with strong backs, stout hearts and a tractor or two. It may be that you will fit in so well with country life that you quickly find *rapport* and make good friends. If so, fine. On the other hand you may miss the earnest talk of organic growing, alternative technology, radical politics, ecology, diet and ley lines which sustained you in the city. If so, you might be better off in a district where a few other freaks have already settled!

No matter where you settle, one tentacle of the System will follow you: the whole of Britain is subject to planning restrictions - some friends, some foe. In a National Park or an Area of Outstanding Natural Beauty you will enjoy some protection from 'improvements' such as motorways, industrial estates and housing developments, but you will find the regulations restricting any new building even more rigorously enforced than elsewhere. This you may regret if you want to extend your living quarters, build more dwellings or add a farm building or two.

One couple whom we know took the supreme chance. In one such area they bought eight acres with no more of a house than a ruined cottage, knowing that planning permission for a new house had been refused. I can understand why they did this. To them there seemed little alternative. They had just enough money to buy such an acreage without a house and be left with a reserve to cover the cost of one they would largely build themselves. Moreover they had first seen these eight acres one summer's day and promptly fallen in love with them. In the autumn they bought a caravan, parked it on the land, fenced off a patch, bought some pigs and hens, and began gardening. They re-submitted building plans - and saw them rejected. Winter dragged on and their spirits sank. With neighbours we tried to cheer them, suggesting they forget the idea of a new house and start rebuilding the ruined cottage instead, stone by stone. At least they would feel they were making progress. As for the bureaucrats, these could be handled in the way I describe in Chapter Eight. This couple's story is still unfolding, but one lesson emerges: do not gamble heavily unless you are equipped to lose, both psychologically and financially.

On the other hand, before deciding on the district - and indeed later when you pick your holding - check carefully that no industrial or residential development is planned which would spoil it for you. In a year's time you don't want to be crossing a motorway every time you milk the goat.

Choosing your own place

How do you find the dot on the map where your destiny lies? You talk and write to others who have already searched or are still doing so, seeking them at meetings, through WWOOF or in columns of magazines such as PRACTICAL SELF-SUFFICIENCY. You study maps and read books about the countryside and about travelling around it. You write to estate agents. After a time you develop the facility to read between the lines of their glowing descriptions, but even before then you will find them a useful way of learning the prices of possible farms; you can then make comparisons between districts and aim for those within your means. You make sorties to likely places - or better still - you take off on a long tour of them, staying awhile in those you fancy and talking to the people who live there - shopkeepers, publicans, estate agents, farmers especially. And if you hear from them of any practising craftsmen or existing self-supporters in the district, seek out such people and learn all you can.

I am a great one for making lists. I did so when we were considering our own place and I recommend the discipline. On one side write down all you like about each possible district, and on the other side all that is bad about it. After a time you will think up headings as a checklist, such as Price, Proximity to Markets, Fellow Freaks, Climate, Soil, Altitude, Latitude and so on. Then make comparisons. Remember that no location will be perfect, because if you seek perfection you will spend the rest of your life looking. Nevertheless, do not ignore any colossal drawback you may discover about a district, no matter how much it otherwise feels right for you. This can hardly be overstressed, for in the end, lists apart, you will probably find that the place or district which *feels* right to you, and the way you intend to live will be the one to draw you to its heart. In proposing that you choose the district first, rather than plumping straight for a likely property, I am suggesting my typically methodical approach to things: more by chance than method, it was the order in which *we* did things. No doubt plenty of people work the other way round - they spot an advertisement for a likely sounding farm, move near to friends already established or simply gravitate to a district they know well and like. No matter what order you choose, you will need some criteria by which to assess any potential property. Perhaps it would be helpful if I mentioned a few.

Your land should be fairly level. If it is too steep you will have difficulty working it, especially for cultivating with a rotovator; if you work on the contour it will 'crab' and if you work up and down hill the wheels will probably slip on the way up, while on the way down the whole thing will be liable to leap out of the ground in a terrifying way whenever you encounter hazards such as a largish stone or hard ground. Moreover, whichever way you work you will have to heave and strain to turn round. Apart from cultivating difficulties, your land will always be at risk when at all bare: especially if is a lightish soil, the run-off from heavy rain may form rivulets and carry the soil with it. Once it has washed into the ditch it is gone for good - along with any precious fertilizers you have been putting on it.

Of course you cannot expect your land to be like a billiard table, and the place would be a dull sort of a farm if it was. A little steep land is nothing to worry about because it can always grow trees and rough pasture; but if it faces north it will not be much use to you. Even quite steep south-facing land will grow useful early pasture, because it absorbs more of the sun's warmth than flat land.

Steep land makes for hard work, so if you are not-so-young beware of too much of it. Slopes mean difficult haymaking - on very steep land it is barely possible. If you intend to run cows or goats, make sure you will have enough meadow land to cut all the hay you will need. If not, you will have to buy in hay, which is costly, or run fewer stock than you may wish.

Look for signs of bad drainage such as rushes, reeds, sedges and flag irises, especially if the land is flat and low lying. Wet land is usually sour, late to warm up, and hard to work, so you cannot expect good crops or pasture from it. You can drain it and lime it of course, but this is all extra trouble and expense which you must take into account.

The best land has a deep layer of top soil over a porous subsoil; it is neither too heavy nor too light, and it is 'in good heart' - that is to say it contains plenty of humus and the essential elements of nitrogen, phosphorous, potassium and calcium, along with trace elements such as magnesium, zinc, copper and boron. Virgin land is usually well endowed with all of these, although it may be acid and need anything up to a couple of tons of lime to the acre to put right - not a cheap operation. An expert eye can tell pretty well whether or not land is in good heart, but if in doubt you can have the soil tested by the Ministry of Agriculture. Any deficiences can then be put right - some more cheaply than others. Take all the advice you can from local farmers and the Ministry.

Weeds - 'plants out of place' - are good indicators of land. Although poor land grows poor weeds, too many perennial weeds on good land will cause you headaches and heartaches for years to come, especially on land intended for vegetables. Bracken, docks, thistles are all difficult to get rid of - evidence of neglect or bad husbandry. But with work and expenditure they will go. All the same, leave some bracken on land unfit for much else, because you will find it useful for the bedding which cows and goats need, and it will save buying straw. Also, since its deep roots bring up minerals which would otherwise be unavailable, it makes excellent compost. Many weeds have their uses. Nettles, which are indicators of fertile land, also make splendid compost, nourishing hay for goats and free, tasty vegetables for people if picked young. However, you will not want *undesirable* weeds in your pasture.

Pasture should have ample clover in it, not only to provide the nitrogen that grasses need for vigorous growth, but to give your livestock the protein-rich diet, either as grazing or hay, which you can ill afford to buy in bags from your merchant. If instead of clover your grasses are having to compete with buttercup, plantain, thistles, docks and suchlike, you will have to set aside a sum to re-seed. Examine your grasses, too, for at least some of your pasture should be high-yielding ryegrass and cocksfoot; the other old-fashioned meadow grasses are fine, for they offer your stock a mixed diet which is good for them; but for *quantity* you will want some ryegrass around, especially at haymaking time.

Take note of the condition of tracks, gates and fences. If there is a long farm track from the road to the house, remember that it is your umbilical cord and you do not want it severed. Heavy rain plays havoc with poorly made tracks, especially if they are steep. Since much of farm work consists in moving things from one place to another, tracks in general should have the firm base that makes them

Look for good grasses such as (L to R) meadow fescue, perennial ryegrass, cocksfoot and timothy

usable in any weather. 'Bad fences make bad neighbours,' they say, so check the boundary fences thoroughly. If you are in sheep country and there is only a double-strand barbed wire fence round the place, they will soon be in to clean up your vegetables. Or if you plan to run sheep, your own will soon be out fraternising with the rest of the district. Shirley and I say that if you are running cows or goats, get rid of all barbed wire fences as quickly as you can. Sooner or later an animal is sure to tear its udder - nature, who would never have created barbed wire, placed udders in exactly the right place for being torn - and then you may finish up with half an animal in effect. I shall be saying more about fencing later - and about gates, for if there are none, or the ones you can see are collapsing, you will have to spend time and money to fix them. Of course, if you intend having no livestock, you only need to worry about the boundary. But if you do opt for livestock and your gates and fences are untrustworthy, your goats will eat your fruit trees, the cows will polish off the cabbages and the hens will have a high old time scratching up your newly sprouted seedlings.

It is pleasant to be, as we are, sheltered in a valley with a warm, phenomenal micro-climate. Above us we hear the wind roar and howl from every quarter, but only when it blows from the south-west does it worry us. And the wind can be a devil, tearing covers from hay stacks and compost heaps, flattening crops and upsetting livestock. So beware of any place unduly exposed to it; trees grown bent by the wind are sure indicators of nature in vengeful mood. However, to be in a valley can mean the problem of frost pockets - as we know. In autumn and spring, when frost does its mischief, cold air flowing down the valley slopes can become trapped at the foot and lie there. You can reduce the risk by clearing undergrowth so that the air flows on to somebody else's place, but in autumn you still may lose your late beans and in spring imperil your marrows, tomatoes and early potatoes.

As you will now know, we were carried away by our surroundings - the view and nearby woodland. If you are going to be materially poor, you will find it less of a nuisance in the midst of beauty: simply watch out that the beauty is not a prime cause of your poverty!

Woodland can be a valuable source of firewood, but on a small property, where good land may be scarce, it can cast excessive shade just where you least want it. It can also harbour more birds, squirrels and rabbits than are good for you; for if yours is the only place nearby growing fruit and vegetables, wild creatures will head for it like city folk to a shoppping complex. Just one squirrel which crawled under the net ate every one of the strawberries we had saved to ripen for our favourite jam last year. As for the view, that in itself may be harmless, but never let distant hills blind you to stones at your feet.

In a seductive way, not unlike the view, the house may be the sugar coating to a bitter pill. Many a cottage of irresistible charm, dripping with roses and described by the estate agent as 'full of possibilities' is, in truth, so full of woodworm, death-watch beetle, dry rot, rising damp, draughts, smoking fireplaces and collapsing floors and roof timbers that there is room for little else. In short, it is a ruin posing as a house. Ours was! This deception matters little if you see through it and make allowance either for a firm of swashbuckling builders to strip it to its bare bones and re-flesh it, or else set aside a year or two for blissful DIY and rent out your land. It will matter enormously, however, if you try to put the house right *and* work on your land simultaneously, for you will age rapidly. We did!

Get a disinterested and qualified person to go through the house – and outbuildings – from top to bottom and give you an estimate of what it will cost to put them right in a way that will prevent their being a lifelong liability, and at the same time which will please you aesthetically. For after all, looks *do* matter.

Make sure not only that outbuildings are sound – see that they exist! You can get carried away by the view, the house, garden and waving pasture and not notice their absence. As I shall try to show, storage is critical on the self-supporter's holding, and you cannot store even a bag of beans without somewhere to stand it. Chances are you will be storing livestock, hay, grain, beans, marrows, pumpkins, root crops, preserves, bottled fruit, wine cheese, eggs, butter, honey, tools, equipment, vehicle, machinery, fuel, fertiliser, seed, timber, firewood and cement, just to name a few items. If the buildings are not there you will have to build them. Without proper storage you will continually suffer losses, and of a kind which can make you immensely sad.

You can be sure that everyone else knows a great deal about any place you may be considering, far more even than the estate agent knows and certainly more than he, or the present owner, is prepared to tell you. When you live in the country you learn that secrets are hard to keep. People know a surprising amount about other people's business. If anyone lacks idiosyncracies, neighbours will soon invent a few; it is quite a friendly gesture, for after all, why should he be deprived of something everyone else has? The problem is how to get people to talk about the place. They are unlikely to tell a stranger things which would harm the owner, a neighbour. At the same time they would dearly love to impart the knowledge burning in the breast. A visit or two to the pub can work wonders; so too can questions asked in a round-about way. With a little practice you will develop your own technique.

So … you have seen place after place, you have made out lists of each one's good and bad points and you know the asking prices. Be careful you are not quoted more than the market price. Estate agents and farmers can spot a romantic fool from the other side of a haystack. He will enjoy a warm welcome. If you *must* pay an inflated price for your dream place, all I ask is that you know what you are buying – warts and all – and that you have the capital to put it right.

When buying and selling property most of us are as vulnerable as children. It is something we may do only once or twice in our lives; we are innocent and ignorant, surrounded by 'experts' who do it all the time and whose chief interest is to make as much money out of the transaction as they can. If we ignore the expert advice of lawyers we risk being victims of the advice offered by estate agents; if we dispense with the expertise of surveyors we throw ourselves at the mercy of architects and builders. At the time Shirley and I moved we were going through our severest 'anti-expert' phase. It also coincided with our 'trust everyone' period. The combination was unfortunate.

Mistakes we made

All went well with the selling of our city apartment. Although here, and throughout, we wisely sought the sound advice of our family lawyer, we resolved to by-pass estate agents if we could. And we did. We advertised in *The Times*, conjured up a procession of would-be buyers and, after one false start, sold the place to our favourite couple within a few weeks for the price we wanted. There *was* the gentleman who came to view late one evening and after phone calls to find out what had happened to his wife, left with promises - and some £50 of cash from Shirley's handy handbag; but anyone who does *anything* takes a risk.

All through, we reminded ourselves that no place is actually sold until exchange of contracts; and so we listened sympathetically to heart-rending stories from really nice people who were *definite buyers* but had to surmount just the one small obstacle of getting a mortgage or selling their own place - or both; but we never let any of them get in the way of the eventual buyer for whom we waited patiently and whose offer was unconditional.

Yes, if the selling stage was a walkover, at the buying operation we were *walked over*. We could not avoid an estate agent for he was in the drama before we were, and he caused no serious problems. It was when we decided to give qualified architect, qualified surveyor and qualified builder a miss that we came unstuck. A local freak, wanting work, knew of this alternative 'architect' who would look after us. We met him on site and quickly found we spoke the same language: experts were extraneous; society was sick; small was beautiful; natural materials were best; one should keep outside the System. So we told him how much we had to spend, how we intended to live, our ideas on converting the cottage, the date we wanted to move in - and left everything to him. Within a couple of weeks he had found a team of 'builders', 'surveyed' the cottage, drawn 'complete' plans, prepared an 'accurate' list of costs which - happy surprise - came to *almost exactly* what we had to spend. 'Go ahead,' we said. And he did.

I am sure there was no intent to take us for a ride. I am sure that he was fully qualified to draw plans and that as far as his costings went they were accurate. But, they didn't go nearly far enough; in fact, they stopped about half way. The other source of trouble was 'the survey'. Only later, when we totted up the cost of putting right all the things that were wrong with the cottage, and which the 'survey' somehow missed, and added the total to the cost of finishing the job - rather than taking it half way - did we understand why the real conversion cost was twice the estimate. Since we have to finish it largely ourselves, we also found that the conversion took about ten times as long!

This error of judgement was every bit as crippling as our other major mistake - buying a property too rough and too steep for a couple of 'oldies'. We had to find the money to finish the house - at least to make it warm, dry and hygienic - and we did so by robbing our alternative technology budget and dipping into our working capital. I spent countless hours building when I could have been earning and farming. As a result the main project of improving the land was put back anything up to a year; and it cost much more because of the short cuts we took in order to make up for lost time.

CHAPTER FOUR

The Intentions

There is a story I enjoy about old George and the vicar. One hot day the vicar stopped to admire his new garden, envious of its flourishing crops and weed-free beds.

Mopping his brow, George explained: 'Sweat and horse manure - that's the secret.'

'No doubt,' said the vicar admonishingly, 'But remember, God's work comes first.'

'I'm not so sure,' came the quick rejoinder. 'You should've seen what a mess the place was when he had it to himself.'

Gardening is of course a joint effort - the work of man and something beyond. Now there are people who say they receive messages about what they should do and when, and I have no doubt they are sincere. It is also possible that the messages have some cosmic origin. There is a group of self-supporters living not far from us, who simply let things work themselves out for much of the time. They could be right. As one of them confided to me the other day. 'Organisation and planning are exactly what we came here to get away from. We believe in doing things when we *feel* like it. It's not just because we're lazy or anything, but because then things work out better - as though we're somehow being guided about what to do and when. To try too hard to go against natural forces would be asking for trouble.'

When it comes to organisation, however, you could probably put me right at the other end of the spectrum. This is partly because, although I often feel very much helped in my work, and deeply *in tune* with nature, guided by a degree of understanding patiently gained over many years, I have never yet been fortunate enough to

1 & 2. The cottage from the rear, then and now. Note rubbish in front garden on left of 'then' picture — also stucco on wall and, in centre, the sky showing through the missing roof.

3 & 4. Friendly bulldozer moves a bit of rubbish to make way for geodisic dome greenhouse.

5. Shirley and Patrick slashing bracken to make way for crops.

6. Patrick reflecting on his solar panels.

7. With Hortensia (at rear), Hannah and Hermione (Hortensia's daughter).

8. Shirley and Buttercup when a few weeks old.

receive any *explicit* messages – and neither has Shirley. It is also because I have been so schooled by previous work to plan ahead and organise that I can barely stop doing so. Perhaps the rather vague guidance that I am now given from time to time would crystallise into something more precise and frequent if I were less of a sceptic, and then I would get out of the organising habit. On the other hand, it could be that I am already being guided by messages from the same cosmic source when I plan and organise. If this is so, then there is less of a yawning chasm between the nearby 'natural forces' group and me than might at first seem.

If I am to pass on the benefit of some of our experiences, I feel quite strongly that I should advise others to devote a good deal of thought to planning and organising, rather than trusting to luck. It may be that in time you yourself will be able to dispense with this *conscious* process and tune in directly to explicit messages, or it may be that you already have such a rare gift. But unless I were sure of it, I would first of all *plan* – and that remains my advice.

Planning ahead

It will be remembered that Shirley and I drew up a five-year plan and telescoped it into three years. How far ahead you plan – whether three years or ten – is not important. What matters is that you do plan and that you change your plan as you go along. You have to be flexible because few things will turn out as you expect. Some pieces of land will yield heavily; some crops will fail; a co-operative produce stall in your nearest town may thrive; your hay barn may blow down in a gale. In the jargon of professional planners this sort of thing is called 'feedback', and they use it constantly to change for the better any plans they make. Since your plan will not be rigid it need not dominate you; but it can give form to your dreams and ideas; it can help you turn them into reality.

If you have made an economic plan to give you some idea of where the money will come from, how it will disappear, and how to allocate your time and energy, this can be the basis for your long term plan. Then, as it takes shape, you will be able to gauge how long to allow for various projects, and you can avoid a welter of mistakes. One of the commonest errors – as I have already stressed – is to take on too many projects at once. Another is to decide on a project and forget that almost every 'down payment' is followed by 'instalments'. You buy a hive of bees, forgetting that from then on you must care for them vigilantly, feeding them, preventing them from swarming, cleaning their hive, trying to cope with disease and predators, and extracting, bottling and selling the honey - if you are lucky. Yet another blunder is to tackle things in the wrong order - such as a decision to buy livestock before organising their living quarters, fencing or feed.

If you have been sensible, you will have bought a place which promises to be right for the kind of lifestyle you hope to live: it will grow what you ask of it. The district will be right for the sale of any surplus produce you envisage and for earning money from other sources - craft, part-time job, seasonal work and so on. These three - lifestyle, land and distict - will enable you to draw up your economic plan, and this in turn will be the basis of a long-term plan which will sort out your priorities, put some costing before you, and attach timescales to your various projects. You can alter this plan at any time, just as soon as you begin to get feed-back.

Perhaps if I trace briefly how Shirley and I have gone through these processes - or failed to - the saga will help you to draw up your own plan.

Our plan materialises

When Shirley and I first thought about living on a little land we had decided that we wanted to live simply, growing most of our own food and keeping all our needs as low as possible, so that our need for income would be correspondingly low; we reckoned that this would lessen our dependence on remote people and reduce the strain that we imposed on the environment. All of this represented the kernel of our philosophy and was one reason why we opted for the 'large garden model' rather than the small farm one. To be reasonably close to friends and family and so that I might earn a modest living as a writer, as I have explained, we felt that it would be helpful to be within a few hours' journey from London where I already had contacts. Bearing in mind our small resources all this limited our choice of district. Since we preferred a region of beauty and one where we already had friends, the Wye Valley was the one we chose.

It so happened that outlets for produce there were also fairly promising, both to nearby long-distance commuters, too busy to grow their own, and to the shops and hotels all around, waxing fat on the tourists attracted by the scenery.

Within this environment we found ourselves conducting our unintentional experiment in reclaiming marginal, derelict land. Our aim for our place was to make it productive and beautiful, and it was important to us that the economic plan which we gradually put together should not conflict with this aim. We changed our economic plan many times. Eventually, as I have described, we saw the plan primarily as asking the land to feed us. We would sell or barter our surplus and use any money so gained to pay, not only for the cost of all food we grew, but for extra food we would have to buy for ourselves and livestock. My writing craft was to pay for all our other needs. We would convert the cottage and develop the land out of capital - our savings, the small legacy and so on - and hopefully keep a small reserve for alternative technology and working capital. Since we had bought a derelict property and, in effect, a ruin, we had two choices: either to work steadily, aiming to make the land fully productive within five years; or to implement a 'crash programme', using heavy machinery and paid help to supplement our own efforts and so achieve our aim in less time - hopefully within three years. We chose the second course.

The operation involved a dozen or more major projects and literally hundreds of small ones. How on earth were we to judge what to do first? By waking up each day and turning our hands to whatever we felt like doing? To wait for 'messages' from the land, crops, animals and buildings to guide us? Or to work out a scheme of priorities so that things got done more or less in the right order and at the right time for crops, livestock, seasons - and ourselves? Yes, we chose *that* course.

We began crystallising our plan in autumn, soon after we had rolled up our sleeves and turned builders. Winter was already breathing bleakly down our necks, adding a sense of urgency to our decisions on priorities. These quickly became clear. They were - and not necessarily in this order: to make the cottage habitable; to make it 'presentable'; to grow a few vegetables on easy land already cleared; to clear more land for vegetables, fruit and pasture; to keep money coming in by writing. Since we could hardly do all of them at once, we had to decide which ones to tackle first. It was easy. Try living in a damp, draughty, unheated cottage at the onset of winter, and *that* priority stands out as plain as the path to the parish church.

The idea of a five-year plan came to us quite soon. To telescope it into three years was an idea which grew only when the twin threats of inflation and food shortage had sunk home. We knew that this telescoping presented a formidable challenge, but we felt more than equal to it. We had come here chiefly to do right by the land. *The land* – as we were to discover time and time again – *must come first.* And so we made it the first priority, to be undertaken when we had made the cottage warm and dry. As for finishing the cottage, that task could be fitted in between others; if necessary over several years. As for money, I would simply have to earn what I could, when I could.

Only a few weeks were to pass before we made our first departure from the plan. While rain beat against the windows all was well; but when calm descended and the late autumn sun rolled the mists of the valley away until the golden trees glistened, and the air was a warm caress, we would invent weak excuses to exchange building trowel for gardening trowel and make furtive excursions down the track or terraces to turn over the warm red earth and plant something. In this desperate way did work begin on clearing and cultivating the flat land by the brook where we intended our main vegetable garden to be.

Once we began to work our land we soon found that its folds and features fashioned the shapes and sizes of our fields, paddocks and gardens. Before very long we grew tired of talking awkwardly about 'that steep bit above the track' or 'the slope behind the cottage where the daffodils grow', and so, in a haphazard sort of way and over time, we gave each piece a name. Thus the fresh land which we first worked became Meadow Garden.

We had decided on the machinery which we would need to buy for most of the work of breaking in and routine cultivating: a rotovator and a trailer which would become part of it, and a power mower. Since we intended to burn principally dead wood we would also need a chain saw. These four items we bought as the need arose. For the tough job of breaking up old, rough pasture to create Meadow Garden however, extra power was called for; so we sought help from a neighbour, Owen, who did a bit of contract work. When he arrived with his powerful tractor and rackety rotovator, we watch with awe and anguish as flaying tynes pulverised pasture into naked earth.

A few days later we were admiring the rich, friable loam of our new Meadow Garden when another neighbour, Cliff, paid us a visit. From previous encounters we knew that Cliff, a kind and rugged countryman of the old school, was characteristically outspoken. He looked around Meadow Garden. 'This won't grow much,' he proclaimed cheerfully,' We asked why. He pointed upwards at the woodland which had attracted us to our place at the outset. 'Too shady for a start,' he said. 'All those damn trees.' We waited. Then he pointed downwards at the soil we had been admiring 'You won't get any sun here from November to March, so the soil will be too wet and cold. Might grow a few summer vegetables on the top half. If you're lucky. Then he smiled. 'But I reckon between us we can work something out of this place.'

Within a few days we had made Departure from Plan Number Two. With Cliff's words ringing in our ears, we marked off a patch on the lower slope of the hillside, just above the track which bordered Meadow Garden. Here the land was not too steep, and well clear of the shade from the woodland. This land would grow the vegetables that Meadow Garden could not. So on the first available day we seized sickles and slashers and set to work, hacking and raking its crop of bramble bushes and coarse weeds and grasses, pitching our weight against crowbars to roll its hefty boulders down to rest at the upper edge of the track. And because this land lay just below the old orchard smothered by bramble thickets, we christened it Orchard Garden.

As autumn died, hazy sunshine gave way to rain and the cottage again enjoyed its share of attention. But the rain revealed another problem: almost over night our track disappeared into a pudding of mud; priority number three swiftly became evident, and with it our first venture into using really heavy machinery. Owen told us of a nearby bulldozer driver who would not only make us a track - right to the top of our land if we wished - but would also use as its base the rubble heap which was spoiling Meadow Garden. Moreover, Owen said that he and a friend would be happy to clear the flat land at the top of the property which the new track would make accessible.

In late October the bulldozer lumbered in. Four days later the whole track was finished. Soon afterwards Owen and friend set to work on the Top Paddock, and priority number four was under way.

Meanwhile as we worked on the cottage, mixing mortar, pointing stonework, humping heavy buckets of wet cement render up and down steps, yanking nails from our secondhand wood, sawing it and hammering it into place, we thought of the coming year. Inevitably, before long, five-year vision took hold of us and we pictured our slopes planted with plums and peaches, apples and pears, strawberries and raspberries, gooseberries and blackcurrants. So, for our new orchard of fruit trees we located a strip which flanked the track from the cottage, ordered a nucleus of fruit trees and bushes from a nearby nursery, and before long we planted them. To replace the forest trees which we would have to lose in carrying out our plan, we planted beeches and poplars too, and transplanted small oaks from any land which we cleared. All these trees would grow in the places already wooded, which we set aside as wilderness.

Right from the day when I had first walked over the property I had had my eye on the plantation of 500 Christmas trees. For one thing many were misshapen and many more were already too tall to be saleable; for another they were on one of our few areas of fairly flat land; and finally I had no special wish to be in the Christmas tree business. Exactly when the idea of felling them in time to catch the Christmas market took shape I am not clear; but by early December I was out among them for three noisy afternoons wielding my new chain saw; and by mid-December they were all on their way to remote places to gladden 500 people's homes With our rapidly developing 'five-year vision' Shirley and I could already see 'The Sanctuary', as it became called - profuse with a handsome show of crops, ripening rewardingly in the summer sun.

Somehow, priority number five had sneaked upon us, and our original plan had clearly undergone a profound change. Instead of concentrating on making the cottage warm and dry, we had become diverted to developing the land; and we found ourselves increasingly concerned with doing both at once. Nevertheless by Christmas most of the basic work on the cottage had been done; the worst of the holes were mortared and the rest were tidily stuffed with screwed-up newspapers. With log fire blazing in the huge hearth and hope burning in our hearts, we celebrated the Winter Solstice with our family in style.

Over the next two years we divided our time and energies among finishing the cottage, developing the land, growing food and earning money from writing. In the first year over 200 people visited us - either idly curious to see 'the aircraft crash' or else eager to help us. Almost as many came in each year that followed. We had no time to be bored.

One of our very first actions had been to get our soil tested - a service offered at low cost by the Ministry, and in our opinion essential. The tests showed that it was basically all right but rather acid and low in phosphorous and magnesium. Over the next two years we spread tons of magnesium limestone and rock phosphate to correct these deficiencies - and, because of our hillside and tiny fields, much of this tonnage we laboriously spread by hand - clambering awkwardly with heavy buckets up, down and across the steep slopes.

In February of the New Year we began clearing the precipitious slope of Old Orchard - as we now called it - hacking away the bramble thickets, pruning the worthwhile trees which were now exposed and felling the badly diseased ones, all to make way for pasture. In March we knocked together sheds made from our secondhand wood, and the place which had been oddly silent came alive with our first goats and hens. Between them they would assure us

The place came alive with out first goats and hens

of the protein in our diet, as well as eventually yielding a useful income from surplus eggs and cheese. The goats had yet another important role: they were ideal for grazing slopes too steep for cows; and they thrived on the initial-regrowth of the brambles and other weeds which we were eradicating.

In April we told the milkman not to call any more. In June we sowed the Top Paddock to kale and swedes. July was an exceptionally eventful month. A builder made us a proper goat house, a milking parlour and a cool room for storing our expected produce. The long job of fencing now began in earnest. At this time thanks to Owen's help with his rotovator, Meadow Garden and Orchard Garden proved their worth: we became self-sufficient in vegetables. And we were able to buy more land - one-and-a-half acres downstream from the cottage, bracken and bramble-covered like the rest, but potentially valuable. Now our cottage would no longer sit precariously on the south-western boundary: but this land had even more significance; with the extra hay and grazing from it we would be able to keep a house cow in addition to our goats; and in time she would give us the year-round cheese and butter which we could not get from the goats alone. By the next month we had cleared the Home Paddock - as this new land came to be called - and in November Owen performed his spectacular and hazardous feat of ploughing it.

The following year saw the momentum maintained. Above the Sanctuary and stretching as far as the Top Paddock was a two-acre, south-facing hillside, too steep to put down to pasture. Our original aim for our place was beauty and productivity - and we meant *every* part of it. With this in mind I recalled how the Buddhists maintain that everyone in his lifetime should plant a tree, and I discussed this with Shirley. We resolved to plant one or two thousand for those who could not or would not plant - for this hillside seemed just the place to make good the deficiency. In February then, we planted with help, the first acre - a thousand Western Red Cedars - to be beautiful in our lifetime, even though their productivity would be strictly for posterity. The same month saw the planting of more fruit trees flanking the track; in March, to help overcome the snag of being in a 'frost pocket' and to add variety to our diet we put up a geodisic 'Solardome' greenhouse; and in the Home Paddock we sowed a herbal pasture ley for grazing; in April we sowed a green manure crop in The Sanctuary - in each case we sowed by hand, and covered the seed by hauling brushwood harrows.

By July we had our solar water heating system working; in November we turned again to the trees, planting more fruit trees and bushes, transplanting all our soft fruit to one selected spot below the cottage - now known as The Fruit Garden; and clearing bramble and bracken from around the young cedars - as well as filling in the gaps left by the hundred or so that had died in an unusually dry summer. That November was the month of another important event in our plan: Buttercup arrived, our week-old pedigree Jersey calf, destined to become our house cow when she calved in two years' time. There was a very good reason why we bought a calf rather than a grown cow - even more of a consideration than price: she would grow up thinking she was a goat, at home on slopes so steep that an unwary adult cow could go tumbling from top to bottom. At least, so we hoped at the time - and our hopes were justified.

The following spring saw the arrival of our bees. Three hives yielded over 190 lbs of honey, thanks to a bumper year and expert help from our neighbour, Harry, for whom nothing was too much trouble.

By then the foundations of our three-year plan were taking shape, though the period had not been without setback and feedback! Two of our original goats did not suit us and had to go: one had agoraphobia, expecting her meals to be brought to her indoors; the other ran away whenever we approached her; the third we loved dearly, so, even though she turned out to be broken winded, we kept her - and she milked well in gratitude. Rabbits lawn-mowered our kale when it was at seedling stage; we grew four tons of swedes, but, despite a potato famine, we could not sell them; so, with the goats, we munched our way through all that we could. To level the greenhouse site we had to shift some two tons of rocks. Rocks were troublesome in clearing the Top Paddock too: with bulldozer, tractor, trailer and bare hands we shifted thousands - ranging in size from golf ball to grand piano. The self-cycling solar water system suffered from friction and air-locks; not until it had been substantially re-plumbed could we get it working.

To get the whole operation going we had to make hundreds of trips and telephone calls - to builders and builder's merchants, machinery dealers and repairers, suppliers of feedstuffs, fencing and other farm requirements, seed merchants, goat and cattle breeders, vets, Ministry of Agriculture departments, the Forestry Commission, neighbours and friends. Hardly a day went by without writing and receiving letters. Hardly a week went by without someone staying with us. It was not unusual to feed eight people at a time - and feed them well, largely from our own land.

Keeping records

All through I was sustained by my *penchant* for organising, and when it came to planning, Shirley had to admit that she could not hold a candle to me. I spent many happy hours at both of them. I think that the processes involved were a kind of daydreaming with the added pay-off that the dreams had a sporting chance of becoming reality. List-making was a wonderful help. As soon as either of us remembered something which had to be done it would be written on a list, to be crossed off when finished. The list would be re-written when it became unreadable. Not only did this enjoyable practice help things to get done, but whenever I was suffering from an attack of 'disinclinitis' it also served as a constructive alternative to action. As another avid list-maker, self-supporter John Manson, puts it 'It's a distraction to actually starting, getting on with, finishing anything. Eases the spirit no end; puts off doing anything a bit longer; or rather lets you get on with what you wanted to do anyway; and having listed the various things you don't want to do, they stop worrying you. For a while anyway. If you aren't a maker of lists congenitally - or constitutionally - I can recommend it. Advanced players - like myself - have been known to put on their lists things which they have just done - and immediately cross them off. Then there's Re-making the List, the Long Term List, the Short Term List, the Field-by-Field List ... the opportunities are endless.'

Shirley and I have found we can get by with three lists:

Farm List: jobs to be done, listed under headings of each field, with a couple of extras: 'Indoors' and 'General'.

Personal List: All jobs apart from farm and garden: which means house, mending, health, correspondence and so on.

Shopping List: a most valuable one if you are serious about saving time and petrol. *Everything* to be bought or returned goes on a master list, for farm, house or personal, large or small. Every time either of us has to go to town we look through it; we write what to buy or return on a small list, which we take with us. With this system, and neighbourly co-operation, we keep shopping expeditions down to once a fortnight or less.

John Manson, we recently learned, shares another sophisticated detail of organising practice with us. In our case it is a stool with a hinged lid that lifts up. John, lacking such a stool, uses a cardboard box. Whatever you use, it must have a sizeable compartment where you can put things as they arrive and which are not in urgent need of attention; such as letters you have read but have not yet found time to answer; bills - unless of course Final Demands or cheery ones offering discounts for prompt payment; useful newspaper cuttings and catalogues. When the lid refuses to shut, we simply pull out a wadge of stuff from the bottom of the pile, deal with the odd thing which still matters, and throw away the rest - which by this time is most of it. We call this vital organisational device 'The Compost Heap'.

Apart from lists, you will be enormously helped by keeping certain records. Here are the ones Shirley and I have retained after discarding others which proved less useful.

A diary: apart from becoming an historical document, invaluable for settling arguments about what happened when.

Photographic record: colour prints of people, big events, livestock - young and mature, 'before and after' shots of development and so on.

Garden maps: record each year the crops grown and fertilizer applied. Help you plan your rotations and keep to them.

Daily egg yield: will help you discover whether you are producing eggs more cheaply than if you bought them. Shirley and I do not bother with daily milk yield which is more fuss than the egg record but it would tell a similar story.

Vehicle log: a mileage record with fuel, oil and servicing expenses similarly will reveal the unpleasant truth of what it costs to run. It may even help to reduce the cost.

Tractor (or rotovator) log: a record of hours worked and when serviced will help it to run more cheaply and last longer. Add fuel consumption and any servicing expenses too if you want to know what it is costing you.

Income: you will need a record of money received and for what, so that you can keep accounts. If you enter items under separate headings - butter, cheese, eggs, fruits, vegetables, honey and so on - you should get some idea of how profitable each enterprise is.

Expenses: similarly all outgoings. Besides noting those necessary for accounting, we also record our food, household expenses (like soap), clothing and any major expenses so that we can not only budget ahead, but also agonise over where all the money has gone. Do try to record income and expenditure from the very beginning.

If you aim to produce a surplus for sale you will have to keep accounts so that you can produce evidence of your true net income for income tax purposes. In Britain, you could be taxed on the value of the produce you consume. The law says that, if you carry on a profitable business which involves producing food, some of which you consume yourself, then when the amount of taxable profit is calculated, you must include the value of the food you have consumed. The accent is on *profitable business*. As a private gardener or keeper of a pet cow or a goat or two you could fill yourself to bursting point without any tax liability on the cost of your produce. Again, running a business making a loss greater than the value of what you consume, you would pay no tax. But if your accounts showed that you were breaking even before allowing for the value of what you consume, then you could be asked to pay tax on it.

You can keep your net income low or nil or show a loss by legitimately charging your farm or garden enterprise with all the expense of running it, including a share of your vehicle and telephone. It is best to group expenses under headings such as Fertilizer, Seed, Feedstuffs, Fuel, Livestock. You can claim depreciation on plant and equipment, including your vehicle of course. If you have an accountant as a friend, it is worth arousing his interest in self-sufficiency - he may do your accounts and tax returns for a sack of potatoes, for a free (working) holiday, or even for love.

I'm told that it is easy to register for V.A.T. and so avoid paying this insidious tax on farm purchases, though we were frightened off doing so, and now regret having delayed it.

Most probably you will feel extra strongly about keeping your tax payments low - if possible at zero - not merely to avoid paying out scarce money but because you disagree with the way so much of the money raised is spent. I know of one farmer who for years steadfastly refused to pay income tax said to be due and vowed he would see any tax inspectors off with his gun if they set foot on his place. Eventually the village policeman was called upon to arrest him. They were friends and so there was no problem in persuading him to go along to the Police Station. The outcome was that he spent a month in prison, with no compulsion to work. This experience he thoroughly enjoyed - the first holiday he had ever had in his life, and at no cost. He still pays no tax. Instead he is taken into custody each year by his friend to spend another holiday - usually around 14 days, for the sentence is based on the sum owed. The only problem is his wife, who has found she is left with all the farm work while he is away. Now, I understand that, since the farm is in their joint names, she is claiming *her* right to have the annual holiday instead.

Short term plans

Once our three-year plan got under way, we produced our first annual plan. This helped us to time our big development projects so that they did not clash with the seasonal peaks of planting, haymaking and harvesting. Now we regularly draw up an annual plan. An important element of this is an estimate of our vegetable requirements, and we generally find out how much to plant by doing the sums backwards. Although we are two, we plan for at least three, assuming the equivalent of one extra mouth to feed throughout the year; this gives us enough for the stream of visitors and helpers we have learned to expect – and welcome. In ample time for ordering, we then calculate our seed requirements, using a table such as the one prepared by the Henry Doubleday Research Association, *Dig for Survival* by Lawrence D. Hills. We draw on our maps the areas we expect to sow to each crop – as far as possible within the plots dictated by our four-year rotation. Once sown, we enter that we have done so in the diary and add the date to the map. We always plan for rather more vegetables than we would be likely to eat, so that if the season turns out badly we shall not go short. If it proves bountiful, with luck we sell the surplus; if not we preserve it or feed it to our livestock. Nothing is ever wasted.

All along we followed the month-by-month gardening guides. No particular one became our 'bible'; we would read several to make sure none had left out some essential task. Each morning now over breakfast we discuss our programme for the day – one of many habits that have stuck. We also try to cultivate – less successfully – another admirable habit: 'early to bed, early to rise'. It seems sensible from many points of view: in winter we would save wood and electricity; in summer we would start work before the heat; and it would unquestionably be a more *natural* rhythm to follow. Now, after the umpteenth lapse we have finally tracked down the basic cause: eating our evening meal too late; and this in turn has been caused by Shirley's understandable reluctance to forsake farm for kitchen. Eventually Shirley has compromised by making a strict rule for herself. Until ten in the morning she works in the house; after that she is out on the land, and any indoor jobs not done by then must wait until dark in winter or till after supper in summer – or else be put off till another day, cheesemaking excepted. As she explains: 'If we make a late start in the morning I only get a little done. The one time I ignore the rule – apart from our 'Special Days' – is when we have helpers. Then I have to spend longer indoors cooking. All the time I have to learn not to be too upset about cloudy windows, dust in the bedroom and unwashed breakfast dishes. Why should I worry what the neighbours think? If you too seek to cultivate this rhythm and you would like to get out of the habit of sitting up talking long after bedtime, you might find our experience helpful.

I would not expect you to take on as much as we did unless you are a decade or two younger, a lot less impatient and with more money behind you. If you set yourself any kind of protracted challenge, however, you will probably find – as we did – that you accomplish more than you dreamed possible in the first year, and progressively less in each year that follows. For our first year we worked about twelve hours a day, seven days a week. We enjoyed it. We grew physically fitter and mentally more alert. But in time we noticed that we were accomplishing less and less. We were slowing down – becoming 'stale'. So we held 'an Extra-ordinary Board Meeting' and passed a resolution to start observing the traditional Sunday off. For the next two or three Sundays we dutifully ploughed through the newspaper and its irrelevant colour magazine, read a book or two, took little strolls, generally sat around – and got thoroughly bored and frustrated. How could we possibly enjoy this forced idleness when the carrots were crying to be thinned or the hay whispered 'turn me'? Either we would have to climb in the van and escape to some place in desperation or else think up some other idea. So we had another meeting and achieved a truly British Compromise which has worked therapeutically ever since. Once a week – on Sundays when possible – we have our 'Special Day'. We get up a little later. We put a cloth on the breakfast table; have coffee and maybe hot, buttered honey buns for a treat. We make no plans, but linger afterwards. We do just what we want to do. It might be nothing, or we might thin those carrots, or do something frivolous such as making a flower bed. One of us might be working while the other loafs. It matters not. Special Day has only three rules: firstly, short of essential routine jobs, like milking the goats and collecting the eggs, and short of some exceptional event, such as getting in the hay before a storm, we do just what we like; secondly, if we *do* work we must not feel virtuous; and thirdly, if we do *not* work we must not feel guilty. Shirley and I have found that in the end we usually work pretty much the same as most days, but what we do hardly feels like work. We look forward to Special Day; it renews us.

We have had to learn the hard way that even though we might tumble the barriers that divide work from leisure so that the two are often indistinguishable, nevertheless, as a kind of regular celebration, there has to be a time set aside for 'play'. In other words some structure in our lives. It has not been easy, for it has carried the taste of the old life and its 'regulation weekends' which we had forsaken. At the same time it has reminded us of the pointlessness of discarding aspects of that way of life simply because they originated in it. In time we have grown to respect the professionalism that has been a prerequisite for success in much of it. We have begun to see that a striving for high standards has its place, whether the activity is pottery, painting, preserving or planting. We have found, paradoxically, that the more seriously we take any activity, the more fun there is to be had from it. So it is with rural self-reliance, and we have reached the firm conclusion that there is no room for dilettantes on the land. You have to know your stuff and apply it; land is too precious, fertility too fragile to be anyone's plaything.

Whether for tree planting, grazing or growing vegetables, land is too precious to be anyone's plaything.

Getting it together is not all professionalism, planning and organisation, however. I would not wish to give that impression. Spontaneous decisions, inspirations and celebrations all matter greatly. While working on most tasks, you will find your thoughts constantly turning to your land and your home. This singlemindedness is a form of simplicity and one of the chief constrasts between rural self-reliance and many other ways of living. It will help to sustain you during the time of adjusting to country life and the realities of living largely on what your land yields.

The adjustments should not be underestimated. Part of the change-over to self-reliance is the throwing away of props and crutches. If you have become dependent on them it is best that they go, though do not discard them too quickly or you may stagger or fall down. What props and crutches? Over-dependence on alcohol and on stimulants such as tea and coffee; luxury food items, invariably over-priced and often over-rich - as well as fancy or imported 'health foods', like brown rice and tahini; self-medication with aspirins, tranquilizers, sleeping pills, vitamin pills and pills in general; the indulgence of being 'off-colour'; many cosmetics, most deodorants, all after-shave preparations; hair dyes; the woman's weekly 'hair-do'; regular entertainment including cinemas, theatres, shopping trips, car drives, spectator sports, weekends away, eating out - these become strictly for 'celebrations'; new things generally such as ornaments, gadgets, 'nick-nacks', furniture, expensive presents, fashionable clothes - for while it is preferable not to turn up at the village fête in filthy rags, there is no need to be overdressed either.

By simplifying our life style we were able to live on less than a fifth of my old salary - last year about £1,000. This kept the two of us and a 'half person' - an allowance for our many helpers and visitors. We spent £180 on food - £72 a head. Running the van (nil depreciation) was £170, N.I. stamps a staggering £127, telephone £97, house rates and insurance £88, clothing £76, treats and outings £71. Sundry items made up the balance. However this was before our house cow calved, and since we spent £77 on cheese, milk, fats and oil, the next year should see food costs cut by at least a third. Remember though, it took us five years to discard the props and crutches which had accounted for so much of our previous expenditure.

It is good to feel the need for such support gently recede. There is a sense of liberation. You become aware that something else is taking its place. And the chances are that this 'something' is the new focus to life which motivates and sustains you - your home and your land, and the agreeable bonds which hold together all those who live on the land. The more that you have become involved in transforming your land - in suffering even - during the time of your planning and adjusting, the greater will be your joys and satisfactions at the harvest.

CHAPTER FIVE

The Homestead

'What have you done to our cottage?' I cried in anguish. Shirley stood by my side, silent with shock, in the space where our living room had been. Above us gaunt rafters arched like rib-bones, open to the summer sky. Around us the walls stood cold and naked in their rough red stones and crumbling mud mortar. Except for one black beam spanning the void, all trace of living room ceiling and bedroom floors had vanished.

The sitting room as we found it – modernised

We were down at the building site on the second of our fortnightly visits to see how the team was progressing. The previous occasion had been exciting; for then the end wall of the living room had been attacked and revealed its secrets; not only a fireplace and chimney big enough to stand in, with ancient bread oven of smooth stones built into it; but a spiral stone staircase alongside. All this had been hidden behind the wallpaper of simulated stone.

The sitting room with secrets revealed - restored

The dwelling was even older than we had believed - almost certainly Tudor. We had left elated, for the team had chipped off the hideous stucco which hid its stone construction, they had taken off the roof's leaky, broken slates, and they were painting its trusses and rafters to protect against woodworm and death watch beetle. Progress, it seemed, was being made. Now however, where there had been a house there stood a ruin. Mike, our 'architect', was unperturbed.

'We'll have it back together in no time,' he said soothingly.

'But the bedroom floors and walls ...'

'They were worm-eaten after all.'

'And the plaster on the inside walls - why has that gone?'

'It fell off. Once we tapped some of it, the rest crumbled away. But don't worry. It's always like this. The worst is over. Next time you're down you'll be delighted. Now, what we propose is this. Let's not re-plaster the walls, but leave natural stone everywhere and point it. It'll look beautiful. I've been thinking about the ceilings in the bedrooms too; when we rebuild we can raise them to show the roof timbers and create a lovely domed appearance. Now suppose we all discuss how we'll re-design the interior, now that we've found this beautiful old staircase?'

We discussed it all - Shirley, and I, Mike and the leader of the team

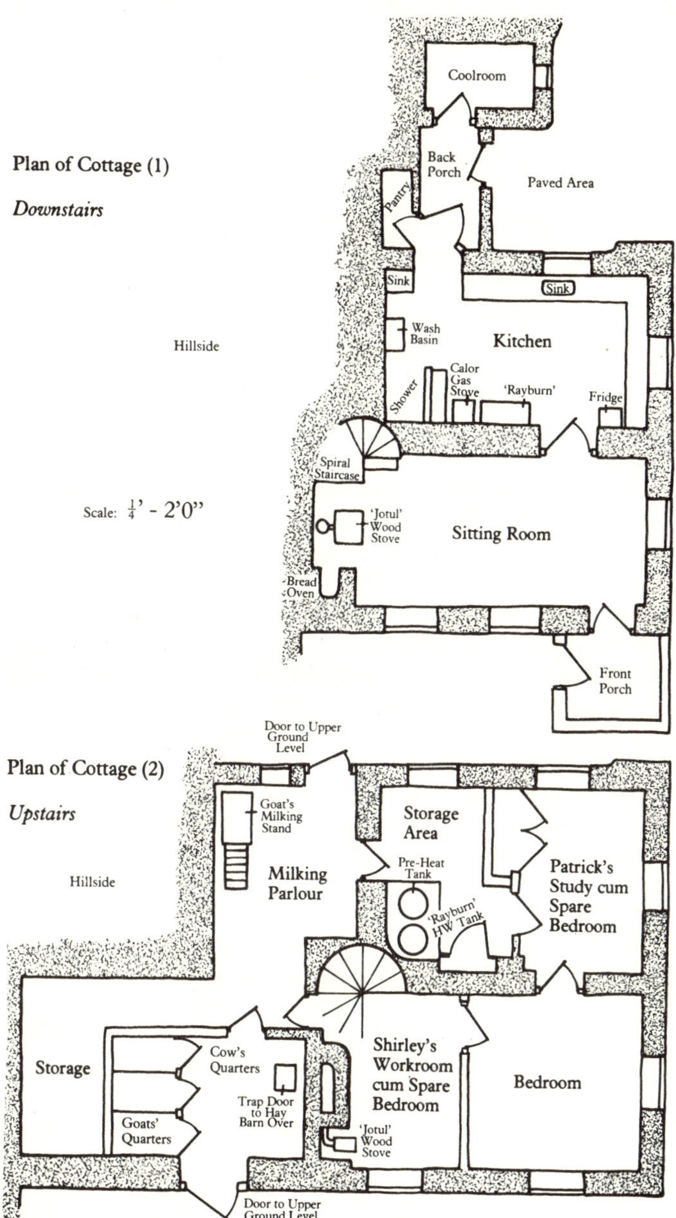
Plan of Cottage (1)
Downstairs
Coolroom
Back Porch
Pantry
Paved Area
Sink
Sink
Wash Basin
Kitchen
Hillside
Calor Gas Stove
'Rayburn'
Shower
Fridge
Spiral Staircase
Scale: $\frac{1}{4}$' - 2'0"
'Jotul' Wood Stove
Sitting Room
Bread Oven
Front Porch
Door to Upper Ground Level
Plan of Cottage (2)
Upstairs
Goat's Milking Stand
Storage Area
Hillside
Milking Parlour
Pre-Heat Tank
Patrick's Study cum Spare Bedroom
'Rayburn' HW Tank
Cow's Quarters
Shirley's Workroom cum Spare Bedroom
Storage
Bedroom
Trap Door to Hay Barn Over
Goats' Quarters
'Jotul' Wood Stove
Door to Upper Ground Level

– a man who we had discovered loved stone the way other people love dogs. And between them they worked such magic on us, that by the end of the visit we had redoubled our enthusiasm for the cottage – despite its present nakedness, it was no longer a hovel but an enchanting home steeped in history; its long-hidden clues were now being revealed.

As we drove away the thought came to me that we should really be living closer at hand so that we could keep a better eye on progress, and Shirley agreed. However, I was writing a book which Shirley was typing, due to the publishers, as it happened, on the day we were scheduled to move into the cottage. The way things were, we had no alternative but to make visits as often as we could spare time from the book.

A couple of weeks later we travelled down for the next visit, our heads brimming with visions of the cottage as it would look. How different from the very first time was the scene that greeted us when we stepped out of the van. Heaps of sand and gravel barred our way. The garage was filled with second-hand wood which we had bought from a demolished barn. Old slates from the roof, rotten floorboards and drums of strange substances filled the tiny front garden. We descended the steep steps expectantly. Chaos greeted us; the back wall had disappeared; rubble was strewn everywhere; the roof was now on, but we still had neither living room ceiling nor bedroom floors; some inside walls had been pointed, but of kitchen chimney there was no sign. The team tried to work the same magic, but this time with strikingly little success. Shirley and I sought solace in the garden and meadow. We walked down the track to the flat land where we would lay out our vegetable garden – and stopped short. There on our best land we were greeted by a heap of bricks, stones, plaster, tiles and rubble several feet high, topped by the modern fireplace ripped from the living room. We looked at each other and we could not speak.

I half walked, half ran back to the house and confronted Mike. It was easily explained, apparently. He reminded me that we had agreed to hiring a mechanical digger and dumper to excavate the back of the site, then explained that he had tried to phone me about where to tip the spoil, but without success. Time was of the essence, so the driver of the dumper had used his discretion.

I recalled our vow that no 'bad vibes' would mar our place and quietly asked him to give thought to how the rubble could be moved. Colouring slightly, he promised he would, and so I concentrated my attention on the cottage. Could we keep to our schedule? Apparently we could. Again I was assured that everything would be all right; by the next visit we would have a cottage once again.

Throughout that summer we continued our fortnightly visits, each time to the same ritual. In an attempt to catch up, we kept on the building team an extra two weeks, after which Mike announced that he would finish the work with a friend or two - at no extra cost to us. The penny had finally dropped! I applied myself to writing and Shirley to typing my manuscript, until a few days before our furniture was scheduled to arrive, when he telephoned to suggest it would be wiser to postpone it. I agreed, asking him to do so for us.

We packed up and drove down the following day, our hearts and our van heavily laden, to a now familiar scene. The cottage which should have been finished was still without front door, back door, rear windows, heat and light. Floors, due to be sanded before furniture arrived, were still rough. Mike and his anxious wife were frantically filling in gaps between hastily erected plasterboard. Somewhere we could hear another person hammering.

We rolled up our sleeves and turned builders - a role we were due to play for the next three months.

Lessons learned

Several lessons can be drawn from this blow-by-blow account. We grew to like Mike and the team. At tea breaks we had friendly, wide-ranging discussions. They may well have said harsh things about us when we were not with them, but no voices were raised in anger by any of us when we were on the site. Yet to place too much trust in strangers can be a chastening mistake.

If you decide to use experts, make sure they are qualified. At every opportunity ask searching questions and make suggestions. An ounce of commonsense is worth a ton of mystique, so never allow yourself to be put off with platitudes. Never have the person responsible for supervising the building job both carry out a survey and prepare an estimate of costs - especially if he knows how much you have to spend. The two figures are too likely to match. Unless you are employing a builder who has given you a binding quotation, add about 50 percent to any estimate of the cost of converting an old house. In any case double the expected time the work will take. Not only is your builder likely to have overlooked costs, not only will unexpected snags turn up, but you too will probably be a culprit, adding ideas as you go along - all of which cost money. In our case the re-design caused by unearthing the spiral staircase undoubtedly added to time and cost. So have reserve funds available, and make arrangements to accommodate yourselves and your furniture during the extended time.

If you can spare the time, do as much of the building work as possible yourself. Shirley and I did not, partly because of impatience, for we longed to begin farming and gardening, but in the end we did our stint of building just the same. However, the more you can do yourself, the greater the satisfaction of living in your new home, the more accomplished and self-reliant you will become. If you simply must employ people, try to supervise the work yourself and try to become one of the team. If that is not possible, at least move so that you live close at hand, and can make daily visits. There are a host of good reasons for doing so, among them the likelihood that you will spot bad workmanship and wrong materials before they are covered up, save useful materials and fittings from being burned, dumped or taken, prevent rubble from being tipped where you do not want it, and - if you are paying the team by the hour and occasionally turn up unexpectedly at starting and finishing time - you will encourage everyone to work full hours.

We chose our team because they were 'alternative'. It was like keeping the work within the family. Only later, when less heavy-handed work needed doing, did we learn that they lacked much needed skills and tools. You can overcome such a problem by being more careful than we were whom you employ. Not that they were wholly to blame for the crippling extra costs or demoralising delay in making the cottage habitable. *They* were not to know about the ancient secret fireplace and staircase. *We* should have resisted the temptation to add 'improvements' after designs had been agreed. And we were *mean* in saddling them with secondhand wood, which demanded much preparation before it could be used - all part of our philosophy of conservation and not buying new, but seen by them no doubt as downright stinginess.

If you want to be conventional and avoid risks, use a proper builder. He may save you a lot of hassle, but remember that he will have loaded his estimate heavily to take care of the unforseen - which you will pay for whether it arises or not. And he will smartly tell you what he thinks if you get 'ecological' and dump secondhand materials on him!

Converting to low-energy living

In the same way that you need aims and a plan for your land, so will you want them for your house. Even if you are sensible or lucky enough to buy a place with a house than needs very little doing to it, some alterations will almost certainly be necessary to turn it into a place fit for practising the art and science of self-reliance. Shirley and I wanted to restore our cottage to something resembling its past, for we are both much influenced by visual beauty and 'vibrations'. This meant the maximum use of natural, local and secondhand materials, a practice I heartily commend. It also involved some discreet, discriminating technology, for one of our primary aims was low-energy living with minimal waste of any kind. This meant insulation in the walls and roof and below the ground floor, slow combustion wood stoves and provision for later inputs of solar heating, for electricity generated from our brook, and possibly for methane.

For insulation we had Mike build into the walls and roof areas 3-inch thick blocks of expanded polystyrene foam, and these were supplemented by vermiculite granules poured into awkward places. We installed two Norwegian Jotul stoves and a secondhand Rayburn cooker for heating the water, and one central heating radiator in my study. For the later energy inputs we left holes in the walls at strategic places where pipes could enter.

Chiefly to avoid wasting fuel, our aim from the outset was to live in very little space. This has suited us well. For easier working, we would have liked a larger kitchen, but because the cottage was originally 'one up, one down' with a similar but smaller extension later added to the side, this was the only place for a kitchen. As part of our space-saving we put in no bathroom, but in an alcove off the kitchen we fitted a shower and washbasin. The other half of the ground floor became the living room, the space above it became our bedroom and Shirley's workroom, and the room over the kitchen was destined to be my study. Both her room and mine now double up as spare bedrooms for visitors. The lavatory was - and still is - an Elsan bucket outside, and we compost the waste from it for the orchard - a wasteful flush lavatory has no place in our plan.

Despite our early tale of woe, after our first winter we have found that we are in fact very lucky, for our cottage has proved warm and dry. Throughout this and subsequent, wetter winters we have stayed wonderfully free from damp - from both the kind that comes down and the kind that comes up! Without our blessed wood stoves no doubt the dreaded damp would have invaded us. For the first three years we kept them going with the dead wood from our own place; after that we teamed with a neighbour to forage under the auspices of the Forestry Commission. In the beginning before inflation made a mockery of *any* figures, we used about £10 worth of electricity a year, but over time our bills have progressively doubled, even though consumption has stayed the same. Looking back, we doubt now whether we would have allowed the cottage to stay connected to the mains supply. We find it uncomfortable to have this unclean energy all around us. We know that although only some 13 per cent of electricity is presently derived from nuclear power, short of a change in public attitude, the percentage will steadily rise as oil and coal threaten to run out. Every unit consumed hastens that day. Central to our whole philosophy is the bringing together of cause and effect, the recognition of the links connecting us all - connecting us inexorably to planet Earth. I will not pretend that we pray for forgiveness every time we flick a switch; but I do find my eyes drawn more longingly to the clean power cascading to waste down the brook just below the cottage. Perhaps if we had denied ourselves the easy way out in the form of mains electricity from the beginning I would have been encouraged to build the small turbine and generator which could light our cottage and run the few little appliances we use in kitchen and workshop. After all, many years ago when we first left the city for the country we had only paraffin lamps and candles, and we got by. If here and now we could be content to use them until the day we can generate our own power, any physical discomfort would be countered by a greater comfort within ourselves.

As it is, the only alternative technology we have embarked on is solar energy, as a way to lessen our use of non-renewable fuel supplies. We installed a system during our second summer. With three square metres of collectors situated below the cottage, we reckoned that hot water should circulate simply by convection - without pump or switches - round some 60 feet of tortuous, lagged plastic pipe, connected to a coil inside a pre-heat tank. This is an extra 35-gallon tank between the mains water input and the conventional hot water tank, heated by the Rayburn. Both tanks we have lagged with a good two inches of glass fibre.

On the first sunny day after installation we kept feeling the pipe where it entered the pre-heat tank. It remained cold. With help we remade all the sharp bends in the system to reduce friction. Still cold ... we must have an airlock problem, we figured. So we forced water up from the bottom of the collector until it overflowed into the header tank. The pipe grew warm, then too hot for comfort! And that evening we had our very first solar showers. Through the long hot summer the system worked well, but with the first hot day of next spring we found the header tank dry. The piping had sprung a leak - inconveniently inside the cavity wall of our coolroom. The airlock was back again; with water unable to circulate, the piping had become hotter and hotter. We remade a whole section to by-pass the coolroom wall, and, by banging a new hole through the main wall of the house, we also eliminated the kink at the top causing the airlock.

All was now well - or nearly. We now found that the system functioned properly only in continuous sunshine such as characterised the previous exceptional summer. To start up quickly in the morning and after cloudy spells we would have to fit a pump to overcome the inertia of 60 feet of water in the piping, as well as a couple of switches to start it and turn it off once the water could circulate naturally. Alternative technology is strictly for the strong!

I would not recommend anyone to become over ambitious about alternative technology in planning for self-reliance. On the face of it, nothing seems more appropriate, but in the present stage of the art it can run away with a lot of time and money; and unless you are rich this time and money could probably be more sensibly applied to your land and your money-earning craft. Our modest excursion cost us around £350. It would have been less if I had made my own panels - more if friends had not been willing to help with the plumbing. Over the years, as the price of electricity continues to double and firewood gets scarcer however I know it will pay.

The kitchen

If the kitchen is the heart of a home, then the stove is the heart of the kitchen. Everything revolves around it. And when in winter, whether for chimney-sweeping or from neglect, it has gone out, the whole house seems to die a little. So when you design your kitchen, design it round your stove. You will want plenty of easily cleaned working surfaces - for baking bread, preparing vegetables, making jam and bottling fruit and for generally cooking and washing up. If you do not have a separate dairy - as we do not - you will also want working surfaces for butter and cheesemaking. A large kitchen table is a priceless asset, provided that you have room. You will need ample storage space too, but do not rely on any low window sills for they will be wanted for bringing on seedlings at various times. So have lots of shelves. There are not many small printed packets of food on our shelves; instead you will find most of them filled with our own produce and many from sources where we buy in bulk. There are rows of jars and bottles, dried beans – haricot, daffa and runner beans – dried peas, wheat grains, wheat flour, oatmeal, dried fruit and brown sugar; not only Indian tea but mint, lime-flower and elderflower 'tea' and dandelion 'coffee'; as well as all manner of herbs and spices. In our pantry and storeroom you will find dozens of jars of bottled fruit and tomatoes, and jar after jar of jams and chutneys. The jars look most handsome. Our storeroom also holds our eggs, hard cheeses, pumpkins and potatoes; soft cheeses you will find in the pantry. Both pantry and storeroom are built into the hillside with inches of insulation in roof and side walls, so that they keep cool through summer.

Until we bought our house-cow-to-be, we neither had a refrigerator nor felt any great desire for one. Through long hot summers nothing, but nothing, went bad. When your food is freshly picked and dug, your milk fresh from goat or cow, and your cheeses are homemade and properly stored, you have little need of a refrigerator. We finally succumbed at the prospect of storing butter, cream and cottage cheese for the longer periods that a house cow's yield of milk would demand.

Some self-supporters swear by the deep-freeze. They say that it releases them from the chore of bottling at the time when all hands are wanted outside for harvesting. We seem to manage without one, probably because we rarely eat meat. Yet that is not the only reason. There is the expense of buying one and running it, and there is the unease of having in the home a bulky piece of fairly high technology which we can do without. Now I know we have a refrigerator; we also have van, rotovator, chain saw and goodies which most of the world can never hope to possess. And so if I say that some of our unease has something to do with collaring more than our just share of the world's resources I lay myself open to criticisms of guilt and hypocricy. I may well suffer from both, but since I have to justify my stance, if I can, I would say that each of us would do well to limit our possessions, especially energy-hungry ones, to criteria based more on need than want. We shall not save the world by doing so, and we shall certainly not achieve an egalitarian one either; but at least we shall have taken a few small steps in the right direction rather than the other way. This unease, however, is not the only one which discourages us from joining the deep-freeze brigade. There is much pleasure in having fresh food when it is in season. We do not want to deprive ourselves of the *anticipation* of strawberries and peas any more than the delight of eating them after long denial. To us there is something essentially wrong about a concept of living in which one can have *all* things *all* year round. It seems a denial of natural rhythms. Without a deep-freeze we eat wonderfully well all winter through, with our stored root crops of beetroots, carrots and potatoes, with dried pulses and bottled fruit and tomatoes, amply supplemented by stored onions, marrows, pumpkins and squashes, and freshly picked leeks, celeriac, cabbages, brussels sprouts and lettuce. Spices and fresh and dried herbs add subtle flavours, piquancy and variety. With milk, cheese and eggs for protein we do not lack.

Once again I lay myself wide open to criticism. Firstly, Shirley and I can manage with the dreaded deep-freeze not only because Shirley is prepared to slave over a hot stove for hours, bottling, but because in our favoured valley we have a long growing season. The farther north and east you go the shorter the growing season; everything ripens at once, the glut becomes THE GLUT, and dodges such as successional plantings are impossible. Moreover, the harder the winter the less chance of growing late or early vegetables; those that the frost misses starving, birds will eat. For such places the freezer assumes more the role of necessity than luxury. And then I talk in one breath of 'natural rhythms' and in the next of eating wonderfully well with bottled this and that *out of season.* Clearly I cannot have it both ways. So what am I driving at? I simply mean that deep-freezing is so easy, there is such a temptation to have *too many* eating pleasures out of season that we would probably succumb and eat peas, and runner beans, strawberries and raspberries every month of the year, and barely distinguish between the frozen and the fresh. With bottling it is different. There is too much untimely, over-hot hard work involved for Shirley ever to get unduly carried away with the amount she puts down; and, with the possible exception of blackcurrants, the end product is not the same as freshly picked. Having said all this, I know that the great 'to freeze or not' debate will continue as heatedly as ever. As a friend pointed out to me: 'It's all a matter of degree, you might say.'

Now I have gone to some lengths to say what *feels* right for Shirley and me in all this. I do not want to be accused of moralising. I do not expect others to do the same as we have done, for each person's circumstances will be different. I can only bring to light principles that seem to make sense because, unlike individual circumstances, they are common to us all. How each person interprets and applies the principles will rightly vary – from outright acceptance, through modification to total rejection!

There are two pieces of kitchen equipment we look upon as invaluable. One is a small blender or vitamiser, which Shirley uses almost daily for turning left-over vegetables into gourmet soups, for making sauces, batters, and mayonnaise, and for fruit flavoured milk drinks. The other is our hand-turned 'Atlas' flour grinder, which grinds, at the rate of a pound in about three minutes, the wheat we buy in bulk. When bread is such an important element of your diet you want nothing less than the best - and the best bread can only be made from wholemeal flour, freshly ground so that the wheat germ has not begun to ferment. Good bread however does not mean hours of kneading. Shirley bakes bread the Doris Grant way without kneading, so that a whole bake only takes about 20 minutes' work.

In aiming for self-reliance, time is of the essence. That is why we do not dry up after washing our dishes and so on. We use plain washing soda, dip in hot water, then stack to dry - far more hygienic than wiping with a cloth anyway. You will save time too if your kitchen has easily cleaned floors. Ours has quarry tiles, laid over a screed above polystyrene foam blocks for insulation.

Food and cooking it

Our way of eating seems right for us and for our little piece of land; but it may not be right for you. We are probably around 80 percent self-sufficient in food, depending on how you do the sums. We could be more, but I am not sure how much that matters. What does matter is that our bulky food should not come from far away and that as little of it as possible should be imported from abroad. In this way we follow the teaching of Ghandi, who urged that staple foods should not cross borders since to do so led to exploitation, corruption and dependence; spices however should, he maintained, for their weight and bulk were small in comparison with their benefits, and they suffered but little from the evils of trading in staple foods.

We would not wish to force our way of eating - nor any other way - upon anybody. We do not believe however that veganism, with its ban not only on eating meat, but also on eating eggs and dairy products, is either the most natural diet or the best one for a self-supporter living in a climate of extremes. Observe the diet of any primitive people and you will discover that even if they do not hunt, they gather grubs, insects, snails, reptiles, eggs and any small animals they can catch. We see nothing inherently wrong with eating a moderate amount of meat, especially if you are doing the heavy and constant physical work which is part of this way of life. We see much that is wrong with eating too much meat; as we do with meat from birds and animals which have been over-fed on imported grains and fishmeal which hungry people could eat; as we do with meat from birds and animals fed on hormones and antibiotics, for we do not want these substances in our own bodies; as indeed we do with animals that have been cruelly reared, housed, transported and slaughtered. Since I for one cannot be sure that most of the meat I am offered satisfies these criteria, I am more comfortable if I do not eat much meat. Shirley feels as I do, but with the difference that she no longer even *likes* meat, and so she eats none at all.

We could of course rear pigs, extra poultry and our male kids and calves; and then I could eat a little of what we did not sell or barter. In this way I would be eating a more 'normal' diet. There are snags however, not just the extra fencing and time a few more stock would demand, but other considerations, as you will see in Chapter Eight.

We have known people eat in strange ways, ignoring both commonsense and reason. We have known of self-supporters - and others - who have become weak and even sick because of their odd diets. One of the strangest notions concerns raw food. Shirley and I are all for eating raw the foods which, for most people, are digestible when eaten raw in reasonable quantities - lettuces, tomatoes, strawberries, apples, sprouted beans and grains, and so on - for they provide the vitamins and roughage everyone requires. But the starch in grains, potatoes and most other root crops is protected by a layer of cellulose which we simply cannot digest. A cow can, for one of her four stomachs is a factory for making bacteria to digest cellulose for her. But we humans cannot. Instead, we cook some of our food so that the cellulose casing round the starch explodes. We all need a certain amount of cellulose - more than people on a high meat, high convenience food, low fresh vegetable and fruit diet consume - for roughage helps food to pass along the gut at the right speed for maximum nourishment. To eat raw, too much of any food which should be cooked however, is merely to waste food by passing it out undigested. Overcooking, that Great British failing, is almost as wasteful, for it not only ruins the texture and flavour of vegetables, but reduces the vitamin content too. Food is too precious to waste by *any* means.

I have mentioned this question of diet because it is pitiful to encounter good people working hard to produce good food, only to fail to obtain its full benefit. In fact I no longer find the subject of diet to be one of endless fascination. Provided that you observe a few basic principles and avoid silly practices, there is ample scope for experiment; obsession with diet however seems to be a stage in awareness through which most of us pass; when anyone becomes stuck in that stage he risks becoming a menace to himself – and a bore to others.

Now eating chiefly food you have grown yourself means that you have to adopt a radically different attitude to preparing and cooking it, the more so if you elect not to eat meat, and more again if – like us – you feel obliged to try not to eat imported food other than small supplements such as spices. For a start you need a lot more imagination. Otherwise your meals will be boring and you may have a mutiny on your hands. Let us take the extreme situation chosen by us. Whereas most meals are based on meat, with vegetables as something of an afterthought, in our household vegetables, grains and beans are the starting point.

In the past a vegetarian meal usually consisted of a 'nut cutlet' or similar with 'gravy' and two veg. and that was that. Today vegetarian cooking has undergone a revolution and all sorts of appetising adventures have become possible, as Shirley is continually demonstrating. It is what she *adds* to our basic foods that makes the difference, especially in winter time. In summer the taste of freshly gathered vegetables, raw or lightly cooked in next to no water, is usually enough – though we do have to keep an eye on our protein intake and eat some cheese or egg, either with the vegetables or separately. In winter and during the 'hungry gap' of early spring, however, Shirley's imagination is let off the leash. It is then that she adds the precise blend of herbs to transform a wholemeal pasta. It is then that a few spices or curry elevate plain beans to dizzy heights of exoticism. It is then that sauces from her hundredweight or more of bottled tomatoes work the same brand of magic on bean casseroles and *quiches*. And the same tomatoes turn up happily in some of her 365 varieties of soup. It is then also that she covers leeks, potatoes, parsnips and artichokes with cheese sauces, flavoured either with stored onions or later their 'bolting' tops.

Winter, too, demands imagination in the salad department, for health requires some raw food each day to keep up the vitamin supply. So she makes cole slaws, and grates carrots or beetroot, preferably with celariac. These she serves with herbs and a mayonnaise or salad dressing. We find that cold, wet or gloomy winter days call for a hot midday meal, and so Shirley serves a soup or, frequently, baked jacket potatoes, either with plain cheese or stuffed with cottage cheese and herbs. Before she makes a soup she invariably looks round the pantry, and if there is something left over from the previous day which could go into it – in it goes, usually via the blender. Soup-making only takes a few minutes but the effects can be shattering. The base for such soups is usually barley, porridge oats, wheat or potatoes, often with milk, skimmed milk or whey added. All good 'outsides' of vegetables are saved for vegetable stock, which also becomes a base for soups. For extra variety she also makes vegetable dishes such as shepherd's pie, pastry crust pies and casseroles. At most meals there will be her own wholemeal bread, and when it is freshly baked a loaf is almost a meal in itself.

Home grown beans are an important staple food for us, either cooked with wheat grains or with flaked oats, or else served with bread. The addition of cereal in some form makes beans far more digestible and nutritious.

Typical menus

Summer:
Breakfast
Sprouted wheat or muesli, fruit, yoghurt, wholemeal bread, butter, honey and jam, tea.

Lunch
Green salad with hard or cottage type cheese, either plain or laced with herbs and garlic; perhaps potato and mixed vegetable salad made with leftovers in mayonnaise, with lots of herbs. Fruit, wholemeal bread.

Supper
Usually a hot meal of potatoes and other vegetables in season, one of which would be served in a cheese or egg sauce. Fruit salad or fresh fruit or a fruit crumble with egg custard or top of the milk.
Herb teas or dandelion coffee.

Winter:
Breakfast
Usually porridge made of oats and sometimes boiled wheat added for a change of texture; milk, brown sugar or molasses; or else an egg; wholemeal bread, butter honey or jam, stored apple if still available, tea.

Lunch
Either wholesome soup of leftovers or of milk with potatoes, leeks or celariac; or else baked jacket potatoes. Cheese, grated raw beetroot or carrot salad; wholemeal bread and butter with honey.

Supper
Bean and mixed vegetable casserole made with bottled tomatoes, Marmite, wheat or oats, served with greens in season. Fruit sponge or pie with egg custard or top of the milk. Or milk pudding.
Herb tea or dandelion coffee.

'Special Day'
Breakfast
Boiled egg, wholemeal bread and butter.
Hot honey buns, fruit in season. Coffee.

Lunch
Salad in season with either leek, tomato or onion *quiche*. Wholemeal bread, butter, honey or jam, fruit in season.

Afternoon tea
Scones with jam or honey, cake.

Supper
Stuffed aubergines, peppers or potatoes, depending on the season, served with any other fresh vegetables.
Fruit meringue or cheesecake or chocolate cake with cream.
Herb tea or dandelion coffee.

Since we began growing our own food, Shirley and I have discovered that our attitude to food has undergone a profound change. When it was something we paid for like any other commodity, while we enjoyed eating, we did so far more casually than now. In our new life, when we see the food on our table, we see also the work of many hands and we glimpse the many miracles which brought it to us. We see too – if we try – the sufferings of those who have too little of it and then we realise our good fortune. We tend to eat more slowly and less greedily, and to give thanks for our food before we eat it. A few years ago, the idea of 'saying grace' would have seemed odd to us, but now it is a perfectly natural pause and an important moment of sharing. Sometimes one of us may express thanks aloud; often we are quiet; sometimes, if the mood is right, we join hands with others round the table for a brief time. We know others who have come to do this too. We see it as one way of appreciating the significance of food – a life force which has generally come to be debased, abused and extracted from the earth with violence. In our home, we try to redress the balance; to eat – so we have found – is to celebrate life a little.

CHAPTER SIX

The Land

Your homestead is the heart of the place where you settle to garden and farm. You will want your dwelling warm in winter, cool in summer, and dry, clean and welcoming at all times. I would urge you to pay your home due regard before applying yourself wholeheartedly to your land; for once you begin to work your land it will demand your love and attention with all the urgency and persistence of an only child. You will ignore it at your peril and at the cost of your peace of mind, for you will quickly learn that your well-being ultimately depends upon the well-being of your soil and all that grows in it - even more than upon the sun, the rain, the snow and the wind which alternately bless and abuse it.

Whether we live in country or city, we ignore the land's well-being at our peril; for in the end human life - indeed *all life* - depends on the health of the land, air and waters. However tempted we may be to pin our faith in mankind's remarkable intelligence, the cosmic intelligence of nature will eventually have the last word. So if you should ill-treat your little piece of land you are merely repeating on a tiny scale the mistake which has brought mankind, and the world he shares, to the brink of disaster - if not beyond. The land asks of you that you understand it and be kind to it.

If you heed its demands you will not spray or dust it with poisons as others do nor will you dose it regularly with chemical fertilizers; you will not grow the same crop on the same land year after year, nor will you over-graze its grassland; you will not leave its soil bare for any longer than is necessary for planting or to clean it of weeds, no matter how 'tidy' its nakedness looks, for if you do it will surely wash or blow away; you will plough, rotovate and dig as little as possible, especially if your soil is 'heavy' and never when it is oversaturated, for any mechanical disturbance carries risks. High among these risks is the harm to earthworms, whose teeming millions cultivate it gently and for free, while helping to make nutrients available for your crops.

You will try to create conditions as close to nature as you can, yet still favouring the species of plant and tree which you select in preference to others. This you will achieve and sustain more effectively by understanding than by brutalising. The chief difference between traditional farming and 'agribusiness' is the gulf between these two approaches.

Certainly we must understand nature, above all we must respect all life, but we must never forget that what we are doing when we are farming and gardening is in essence unnatural. This is far from being a totally non-exploitive way to live, for we use our human cunning, knowledge and technology to exploit animals, birds and bees the whole time. Up to a point this may be all right, for so strong is the reproductive urge that all life forms produce a surplus to take care of predators, disease and mischance. Thanks to this there are excess calves, eggs, milk, grain, fruit and vegetables, and we are perfectly entitled to live on them - just as other species do. We go wrong when we are greedy and thoughtless. If we take all or too much, we rob other creatures who also need to live, and not all of them are our foes by any means. Worse still, if we use brutal means such as herbicides and pesticides indiscriminately, we kill our friends as well as our foes.

Farming and gardening *are* unnatural because we breed varieties to suit *us* rather than the species with which we interfere; we select for yield, taste, colour, ability to grow in a cool climate, and so on. Then, predictably, the species succumbs to predators or climate. And we interfere with the processes of natural selection too when we try to eliminate the favoured species' competitors. There is no turning back, of course - at least not unless we are prepared to return to our hunter-gatherer existence in more favoured climates. Instead we must stop being pious about 'natural methods' of growing, recognise our immense power over other species, and use it as sensibly as we can for good in preference to evil.

The anatomy of soil

Soil is rock, pulverized by the elements into powder of small or large particles and leavened by humus - decayed or decaying organic material which once lived as vegetation, animal, insect or other creatures. Many of the nutrients that plants need come directly from the powdered rock; some of them are from the rock *in situ*; some have been blown by the wind or deposited by glaciers or rivers; others have been brought in as the excrement or bodies of birds and animals; possibly some also as a constant 'bombardment' from outer space.

Gentle acids in rain water combine with the activities of countless millions of soil bacteria to act upon soil particles and so release a constant supply of plant foods. All this determines how fertile soil is. Any land which is neither over-farmed nor over-grazed nor subjected to leaching from over-prolific rains, grows richer and richer in plant foods, and so more fertile. If it is farmed with understanding, fertility can still be maintained or even improved: the farmer simply has to take care that the rate at which nutrients are exported to towns and cities does not excede the rate at which they are replaced naturally. If, however, land is farmed greedily as if it were a factory floor, then nature cannot keep pace and the farmer is forced to apply quick-acting chemical fertilizers in an attempt to replace lost nutrients. This may sound all right but it is not. A little 'artificial' fertilizer at the right time may not do much harm, and even if it does, the injury soon heals; but repeated applications upset the ecology of the soil: earthworms and bacteria die, humus is lost, light soil becomes sandier and heavy soil more clayey; until eventually the farmer resorts to feeding his crops, not through the soil, but 'intravenously', so that soil becomes merely stuff to stop crops from blowing away or falling over. He and his land become 'hooked' on 'artificials', totally dependent on the vast companies that make them, and susceptible to each year's swingeing increases in fertilizer costs. In consequence he finds himself at the mercy of his bank manager, anxiously watching the rise and fall of interest rates as he plunges deeper into debt. In place of his role as universal provider of sustenance he finds himself a mere rural appendage to the dominant, city-based industrial system.

There are many types of soil and each will respond differently to cultivation, to fertilizers and to the kinds of crops grown. If soil particles are large, as in sand, the soil is said to be 'light' and it is easy to work in most weathers, though it will probably be 'hungry' - nutrients will wash through faster. If the particles are small it is 'heavy' and harder to work, but nutrients are more likely to stay put. 'Light' and 'heavy' in this context have nothing to do with the weight of the soil. The wise, small farmer or gardener soon gets to know the vagaries of each piece of his land: which crops grow best and where;

which field or plot is early or late; which is over-drained, which is boggy - and so on. When too many weeds creep in he does not spray indiscriminately with poisons so that friend and foe are killed alike; instead he smothers them with a green crop or with pasture, cultivates judiciously or gets out his hoes. If he is wise he considers not only his actions, but the effects of them, both short and long term. So it is with all the other problems which are inevitable as soon as he interferes with nature by adapting her to his needs.

He remembers that he is but part of a chain of life - a food chain - and that to break this chain at any point is to weaken its whole. And since he is dependent on it, his own health and security are correspondingly weakened. At the bottom of the chain are the countless millions of bacteria, most of which live in the soil, many with the unique ability to absorb the essential element of nitrogen direct from the air. A single speck of soil contains millions of benign bacteria. Next above them in the chain are countless microscopic grubs, insects and other creatures which feed on these bacteria; and upon their residues live the vegetation on which many birds, animals and man feed. Some are herbivores, feeding solely on vegetation; some higher up the food chain are carnivores feeding on the herbivores; and some lucky ones - like man - are omnivores, capable of feeding on both.

The strength of the chain lies in the natural processes of birth, death, decay and rebirth. If these are interfered with by poisoning or burning, then life is lost at some point in the chain and the rest of it is weakened. There are less apparent ways of interfering with the chain, too, and again man is the offender. A classic example is his hunting of animals which prey on herbivores. When he reduces their numbers, vegetation is overgrazed, the soil is bared, it blows away and deserts result. The goat has been called the creator of deserts, but this is unfair. It is man who has created deserts either by removing the goats' predators or by placing goats where there were no predators - as on remote islands to provide food for shipwrecked sailors. Sometimes, rather than a chain, the analogy of a pyramid is used, since numbers of creatures are large at the bottom, fewer at the top.

Traditional farming

On his little patch of land the self-supporter can gain comfort and sustenance by creating the right conditions for preserving the chain of life, no matter what crimes are committed beyond his boundaries. For example he can imitate traditional mixed farming methods where the land feeds crops and pasture, which feed animals and people, who in turn feed the land with their dung. The farmer and his family are part of the cycle. They *are* their farm. By rotating crops and animals around his land, the traditional farmer kept at bay the build-up of parasites, weeds and diseases which now plague the specialist in 'agribusiness' who indulges in monoculture and must apply poisons to his land in a desperate, unending battle. It is unending because the scientists, into whose hands he has placed himself, are barely able to keep one jump ahead of the weeds, pests and diseases they fight; just as they claim each victory, so the enemy adapts itself to resist the newest poison and re-emerge for the next round.

The traditionalist accepts that *some* losses are inevitable, that birds and bugs are entitled to a fair share of his crops. He does not demand that his land be as free of weeds as the Centre Court at Wimbledon. He controls his losses by a blend of cunning and gentleness. When he rotates his animals around the farm he prevents the build-up of parasites such as intestinal worms, for the enemy of one species is harmless to another. By rotating his crops he similarly controls pests, diseases and fungi and either smothers weeds with vigorous pasture and green manure crops, or 'cleans' the land with row crops which allow hoeing and cultivation.

The factory farmer keeps his livestock caged up in cruelty, and pays dearly for buying in feed and often wasting dung. The traditionalist tries to run his stock naturally over his land so that they can feed themselves and drop their dung usefully, so that he is saved both bought-in fertilizer and the cost of physically spreading it.

Although traditional farming is adaptable to a host of varieties in practice, the principles remain the same. It is possible to farm and garden without animals. In some ways it is easier, in others harder. Apart from the boundary fence, you will not need fences. Vegetables are less troublesome than animals - they rarely get out of place to wreak havoc in a few minutes. But you will miss the benefit of animal manure - unless you expensively buy it in. And to maintain fertility you will have to rely more heavily on resting your land, with green crops eventually turned in, and on making and carting vast quantities of compost, which is one of the most laborious jobs in the game.

Whether or not or not you keep livestock will depend on the kind of person you are - whether you *like* animals, whether or not you eat meat, eggs and milk products - and how much land you are lucky enough to have. As we have seen, it can be a mistake to keep animals on too little land.

It can also be a mistake to try growing unsuitable crops. Wheat is a

The laborious job of making compost
from thick bracken

classic example. Wheat prefers a fairly heavy soil and a sunny climate to ripen it. If you grow wheat on light land you will probably get a light crop, which may be all right if you are not greedy, but you will have to replenish your land by resting afterwards or by applying heavy dressings of manure or compost - or by all three. If you attempt to grow it in too wet a climate you will encounter all manner of troubles in harvesting.

For a number of reasons Shirley and I do not grow wheat. We do not have enough arable land, our soil is just a little too light for comfort, our climate a little too moist, and no matter what John Seymour says about threshing it over the back of a chair, by all accounts the whole harvesting process on a small scale is a lot of bother, and we are already overstretched. For the first couple of years we bought our wheat in half-hundredweight bags from a good commercial organic grower; then we looked around for a nearby self-supporter with the right growing conditions, at first to buy, later to barter.

Remember one thing: farming and gardening without chemicals mean hard work. To select the right companion plant to keep away predators, or to encourage an ally to eat them, requires more intelligence and cunning than buying some nasty stuff to squirt around. To hump heavy compost or rotting dung up and down hills and over soft ground and then chuck it around, will make you more puffed than using handy plastic bags of something concentrated which spreads easily. Shirley and I prefer the hard way, not because we are masochists, but because we believe it grows tastier crops which are better for us, because we can see the cost of chemicals spiralling out of control, and because we think that in 20 years time our land will be even more fertile than it is today, whereas long before then other land will not.

I have already explained that we occasionally 'sin'. This does not worry us unduly, for the whole subject of purity is fraught with paradoxes. We find ourselves smiling a little at the person who spins and weaves his own clothes yet runs a car; as we do at the vegan eating foreign nuts and fruit grown by exploited labour and imported in polluting ships; as we do at the meditator whose personal growth is at others' expense. We do not aim for absolute purity. We believe in being discriminating in our use of technology rather than accepting *the lot*, for all of us are where we are, and we cannot put the clock back. What we *can* do however is reverse a few patently suicidal trends, live a little more simply, become less totally dependent, involve ourselves actively to achieve change for the better, and try to set an example.

When you make your economic plan and your long term plan, consider carefully the crops and livestock you will grow. Take into account your own preferences, the capacity of your land, the suitability of your climate and your proximity to markets. Seek the advice of others - not just books and experts, but local people who know the district. Gradually you will select a mixture of vegetables, cereals, fruit and other trees, and of birds and animals which will demand year-round care if they are to reward you. Gradually you will divide up your land into fields and plots which make the best use of the variations in soil, slope, aspect, sun, shade, availability of water, nearness to house and buildings, existing tracks, fences and gates.

As soon as possible decide on a crop rotation plan, for your vegetable garden and for your pasture and arable land if your place falls into the 'small farm' category. Most probably you will choose a four year rotation for both. Our place is too steep and small, its fields plainly unsuitable for rotating crops and pasture systematically, but we have imposed a rigid four-year rotation on the vegetable garden. We only wish we had done so from the word 'go'. Because we were planting as we cleared and were generally disorganised we had rows of this and that everywhere for the first couple of years. If you land yourself with such a muddle you will have the wrong crops sticking out like islands in the middle of the wrong plot, once you have established what should grow where in each of your rotation plots. To avoid this sort of trouble is all a matter of getting organised over time. If you can do this you should be able to accomplish most of what you plan despite those three basic constraints - the time, stamina and money you are able to muster and use to the full.

CHAPTER SEVEN

The People

Our dream of living on a little land has not unfolded in the way we expected. The structure and tempo of our days are a constant surprise to us, and if the reality has proved less idyllic than we imagined, at least we have almost never been bored. Nor have we felt that our time has been wasted.

People are the chief reason for the difference. People enter our daily life in rarely interrupted procession. People are one of the principal differences between our new life in the country and our old one in the city. Out here we see fewer people, for we are not constantly brushing against strangers in streets and stores, buses and trains. And because so many encounters are with people we have met before, we enjoy more opportunity to *know* them. The postman not only delivers our mail, he collects it, and we usually have something to say to each other. Young Tom, who keeps my machinery running, is getting interested in solar heating and I can offer a few words of advice - as indeed he can when we have any problems with our home-made wine. Old Frank, who has gardened for a lifetime, and has come to help us out a few times, was shy at first, treating me like a boss; but over cups of tea and talk of roses and shallots he has taught us far more than gardening. Perhaps it was shyness too that made the quiet and diminutive Roberts sisters seem aloof. Soon after we arrived I approached them - possibly too brusquely - about buying some of their land which adjoined ours. Over the ensuing months we talked amicably about their two house cows and country matters generally on any occasion when we met, but if I steered the conversation towards buying the land the answer was always a polite 'no'. Not until we acquired our first goats and the sisters had been over to tea were they probably satisfied that we were not speculators tarting up a cottage for a quick sale, but true neighbours. And not until then did the answer change to a polite 'yes'.

You cannot *insert* yourself into a country community any more than you can buy your way in. You must be invited, and the invitation will usually be made obliquely and a little at a time, so that either party can retreat in a seemly way if a mistake has been made. To enter you must earn respect. Now this is not won by wit and charm nor by any kind of display, but by repaying kindness with kindness and by other deeds, which matter more than words - among them the same kind of hard work that country people do. For if you are to be part of the country you must work in it - on the land, at a craft or in some way that benefits the community. You can never be part of the country if you commute to a biggish town or city, nor if you retire to a district where you have no roots, for you will be seen as an outsider; and then if you cannot work as younger people do, you will have to earn respect in other ways.

For a long while after we arrived in our district we had that feeling of being watched. More than anything else, I think it was our capacity for work which earned us any place in the community that we may now enjoy. Work has to be done professionally to be respected. If a self-supporter is to gain acceptance by the community he must cross two barriers: firstly he is an outsider; secondly, because of lifestyle and scale of farming or gardening, he is something of a freak. And if he is bearded or bedraggled as well he will not only look different from the rest of the district, he will *feel* different too - and that is not always helpful. A newcomer must not act superior nor in any patronising way. Because the tempo is slower, because most people know each other, because each member of the community sees fewer people than the average city dweller, relationships are bound to be different. And so, if you are to avoid the blunders which will delay your acceptance, you would do well to be perceptive, sensitive and observant.

Blunders are easy to make. Early on I asked our neighbour, Cliff, if he would help me put up a cupboard. He readily obliged, and when the job was done I asked how much I owed him for I knew that his time was precious. 'You owe me nothing', he snapped, and walked off in a huff. Humbled, I realised that I still had to learn that in the country, people are people first and functions second. In the city the order tends to be reversed; you do not expect friendship from the man who punches your ticket nor he from you; no more than from policeman, milkman, car mechanic, or street sweeper. This is not surprising, for city life has a surfeit of people, a shortage of time and an element of chronic suspicion. In the district where we used to live you would be foolhardy to leave your car unlocked in the street for five minutes; here a vehicle is safe all day and probably all night - and people rarely lock their houses. In city conversation it is imperative to come quickly to the point; here impatience can lead to a blunder. Before you get to The Point there is the weather to discuss, crops to compare, family to ask after, who is up to what, the price of things, bureaucrats and other ailments. The other evening when a neighbour telephoned, it took him more than ten minutes of costly telephone time before he could get to the embarrassing point about a boundary gate I should have fixed, where one of his distant fields adjoined my land.

Now these matters may not *seem* important but they are. You will find that you depend on other people more than you suspect, and in time you will probably be able to help them too; but the two-way flow cannot begin until you have been accepted. When it does, most likely you will be astonished by it. On that unforgettable day when we arrived in our laden van to find our cottage uninhabitable, a virtual stranger gave us hospitality and continuing moral and practical support. Since then neighbours have helped us build, lent us equipment, given advice, milked our goats, shopped for us, entertained us - even risked life and limb for us. And gradually we are finding that we can go some way to redressing the balance.

Tomorrow will look after itself

Help is what it is about - and the word makes a nonsense of 'complete self-sufficiency'. Trust is what it is about - and the word has to re-enter one's vocabulary. Faith is what it is about - and faith is something to clasp to your breast when fears, doubts and setbacks threaten to eclipse your dream.

Shirley and I were slow to grasp the full significance of all three - I was especially slow and I still have much to learn. This much however we have discovered: that if you are doing what *feels* right - right for you and right for something greater than yourself - there is a strong possibility that help will flow to you, and that coincidences will punctuate your life, along with other happenings, which defy rational explanation. You have to expect the inexplicable to happen, and yet not wait for it passively. Rather you must go forth to meet it, working for it, while all the time keeping aware - recognising opportunity and taking chances.

Early in our first spring here, when the slough of winter was still upon us, I reached a point close to despair, for we did not seem equal to the task we had set ourselves. Then at the critical moment, two strong American students came into our lives from the blue, stayed first a day, then a week, then two weeks, choosing to help us rather than continue their walking holiday. Together they accomplished the impossible: they shifted mountains of scrap, dug ground and cleared scrub. When they left they did so grateful and refreshed, and - so they told us - with a new outlook on work, on the direction of their lives.

From then a floodgate was opened for us. Help flowed our way. When our new solar heating refused to work, a practical physicist materialised and all but remade the system; when a pile of huge rocks fell on our vegetable terraces, five hefty men appeared and heaved them away. The very week when we were ready for our Jersey calf a

Five hefty young men moved rocks from the garden

top breeder telephoned, offering us one which we promptly bought - and it was much the same when we were ready for goats and, later, bees; when our income has run dry, work has appeared; and time and again when our spirits have sunk, someone has arrived, phoned or written to inspire us.

The buffeting of the elements, day-to-day mishaps, even long drawn-out adversity - all these you learn to expect, for they are as tests for your selflessness and determination; but the timeliness and aptness of the help which can flow to you and surpass them cannot readily be explained.

When the significance of The Flow dawned on us, as yet we had no name for it. We learned of its name from a friend who had observed it at work when he was a member of a commune. Chief among the worries of the commune was a recurring lack of money, yet they gradually noticed that repeatedly, just when the kitty was all but empty, something would turn up – it might be work, a gift, a debt cancelled or a chance to barter.

The Flow is very important if you are to succeed in living on a little land. For it to work you have to believe in it. To experience it is to find release from fear and anxiety; in their stead, hope and strength will blossom. It is nothing new. Remember, Jesus urged us to put away anxious thoughts about food and drink to keep us alive and clothes to cover our bodies. 'Is there a man of you who by anxious thought can add a foot to his height?' he asked. 'So do not be anxious about tomorrow; tomorrow will look after itself.'

Working weekends on farms

Some years ago a lass called Sue Coppard started a small revolution. Sue wanted to escape from the restrictive boredom of weekends in the city, not merely to flop, but to balance the work of her head with the work of her hands. The snag was that she knew no farmers. Now Sue is not one for passive acceptance of situations, and she soon found a way out. She started an organisation – Working Weekends on Organic Farms, and advertised for members willing to work, not for money, but for food, lodging and instruction. The idea caught on and today WWOOF – as it is known – is flourishing.

Shirley and I could never have achieved what we have without the help of the scores of 'wwoofers' who have come to us – most of them for weekends, some for stretches of several weeks. Many come to us again and again, and we have made a host of good friends. 'Wwoofers' are mostly young people. They pay – at present – £2.60 a year to be members and for this they receive a quarterly newsletter which lists member farms and weekends scheduled so that they may have the company of two or three others.

WWOOF is an odd phenomenon. On Friday evenings, straight from a week of work, its members willingly undertake long journeys – often hitch-hiking – to strange destinations and weekends of work among strangers. What is it that sets these people apart from others so that they spend their weekends, and even holidays, working without the inducement of money? Obviously they share a feeling for the countryside as well as a common interest in growing food, and often in husbanding animals. But this is not all. I can recall none who have not shared our views on living more simply, and on the need to bring about a gentle revolution by setting an example. Their views set them apart, and doubtless many seek the company of others of like mind; for living with such views in a society committed to industry and growth can make anyone feel lonely. Some are livelier than others; some are more troubled than others; some work harder than others; some ask more questions than others - indeed some ask none; some seek to be part of the household while others stay aloof. Yet none so far has failed to help us, and nearly all, we suspect, have gone away a little more enlightened and more hopeful. On the Friday evenings of WWOOF weekends Shirley and I invariably experience a mixture of apprehension and expectation, knowing that presently we shall have in our home two or three total strangers, themselves no doubt feeling something of the same. Sometimes we greet each other shyly; sometimes there is immediate *rapport* and a surge of joy. It is a strange period, not without strain. But the strangest part of it all is how we discover, time and again, that these unknown people are not strangers, for so strong are the ideals we share with them that we feel we have known them before.

A few months before you are ready to work your land I would recommend that you consider joining WWOOF, (the current address is in the Source Guide.) If you do, and both parties are to reap the utmost benefit, the fruits of our experience may help. We would say: encourage your helpers to feel part of the place - equal partners during their stay; over breakfast discuss with them your plans for the day; explain carefully each job you ask of them, for many will be

Much needed help at potato harvesting

inexperienced and – you will learn – some who profess to be are not; after they have begun a job, correct any faults; stress the value of tools so that spades and hoes are not left in the grass to rust or disappear for ever, nor break through misuse, nor are put away uncleaned; avoid monotony by introducing variety – a couple of hours at a stretch on one job is usually enough for uninitiated volunteers; and at times such as potato planting and haymaking, when one job must predominate, a few words of explanation will create understanding; feed them well – they deserve it; where possible, talk with them while working; in the evenings draw out the shy ones; and thank them.

Besides helping both farmer and learner, WWOOF plays a small, but valuable role in bridging the gap between city and countryside and reviving interest in the land. It is one way in which people can take the first steps to getting started. It needs every encouragement and should grow.

The visitor problem

'Wwoofers' who are usually strangers are nearly always welcome; visitors, who are often friends – it must be admitted – frequently are not. We have planned our day. I may be writing; Shirley may be in the middle of a big bake; or we may both be working outside, racing to beat a change in the weather, when sudden voices intrude. Children scamper across a seedbed; a dog scares the daylights out of our cat; and we are expected to down tools and slake thirsts; for, in many a city mind, the country is a place of play where lucky people with money grow food for free in idyllic surroundings. Sometimes I wonder how I would be received if I were to burst in upon *them* in my working clothes when they are negotiating a vital contract, entertaining the boss and his wife or removing an appendix.

Of course we welcome visitors – friends most of all. We are acutely conscious of our good fortune in having land and we want to share it with others. We need the affection and stimulation they bring, and we are grateful that they *come to us* and save us the horrors of travelling far. But in all conscience we plead that they write or phone beforehand and are understanding if we ask that a visit be postponed awhile. For what each visitor fails to appreciate is his lack of uniqueness; we may have entertained six people the day before; and we may have received 200 over the past year; which is a formidable amount of time, talk, meals and cups of tea - and one can have too much of a good thing.

After six months of 'self-sufficiency', Margaret Findlay in Somerset summed up one aspect of the problem neatly. On half their previous income, with inflation gnawing at their vitals, the Findlays soon found food to be a bone of contention. 'Visitors come in an endless stream - all truly welcome,' she stressed, 'But they do eat, and there's nothing like a farm to sharpen the appetite. Brown home-baked bread and home-made cheese is quite good enough for us, washed down with home-brewed ginger beer, but it just won't do for them. And even if it would do, they are consuming our next week's rations,' She admitted that it sounds mean to ask for two pounds a day for extra food, light, heat and laundry bills, but confided that it was 'a sad reality'. Added to which, entertaining meant taking time off - 'In the words of the industrialist, lost production'. Yet if she waited four or five years till they were established, they would be forgotten, she feared. And she asked 'How does the rest of mankind on the breadline tell its best friends?'

As Margaret Findlay's experience highlights, visitors are a mixed blessing. Some are friends, some acquaintances, some strangers; some come that each may enjoy the other's company, some to learn, some to work, and a few as if they were taking the kids on a trip to the zoo or worse still - to see 'the aircraft crash'. From the outset Shirley and I resolved never to turn anyone away unless the conditions prevailing would make the visit unfruitful both for visitors and ourselves. Few have been put off. Now, however, we have had to ask that those able to work with us should do so for a while, depending on length of stay, and we choose tasks that will allow us to talk as we work with them. We drop a gentle hint that those who can afford to might bring a gift of food as a treat. And we discourage anyone from turning up without warning.

All this does help, but it is a fact of life that not all problems are soluble, tidy though life would be if they were. The visitor problem is in that category; and so, when you have your own place you should be prepared for the dilemma. And you may feel, as we sometimes do that where food is concerned, whether you entertain lavishly or frugally, you cannot win. If you serve your usual plain fare, you can see the hurt on their faces that you didn't think them worth 'putting on a show'; yet bake a special cake, save up three days' eggs and pluck your only greenhouse aubergines, and you'll be greeted with something like 'All right, this self-sufficiency - living like lords'.

Sooner or later you will come across a special breed of people - officialdom. Many are sympathetic and generally helpful, but the fact must be faced that we self-supporters are not very popular with some of them. Our farms are not big enough to be 'viable'. We have this habit of letting people sleep in barns without first seeking 'planning permission'. We keep goats, which are simply not recognised except as a giggle. If we live in a community we are probably communists. We cannot be exactly placed, for we are poor and scruffy yet intelligent - and if we are intelligent why are we not in 'a good job' like they are? In short we represent something between a threat and a bad joke.

On the whole, the best plan is to avoid confrontation. I know of one farmer who was denied planning permission to build a house, but discovered he could build a stable, in solid stone. So he did - complete with kitchen and bathroom. He was very fond of horses ... It is essential to avoid what Shirley and I did about our pond. Down on our meadow we have a boggy patch near the brook, fed by a spring, and the land is no good for anything. But it would make a nice site for a pond, so Shirley said one bright summer morning. It would look pretty, it would attract wildlife, and we could do with the home-grown protein of the shoals of carp we could rear in it. I agreed. Now, possibly thanks to The Flow, a bulldozer was working on the next property, and when the driver saw the job he said he could do it the very next week. For a modest sum. All he needed to know was the proper depth for carp. As a starting point Shirley phoned our nearest university, who put her onto someone else, who put her onto a man who said he would come out. He did - the very next day, complete with collar and tie. He knew about fish, he told us, but not about ponds. However he would put us on to someone who did. A couple of days later another Collar and Tie turned up, complete with slide rule. He told us everything we wanted to know except for the forms that we must fill up in sextuplicate with a diagram, and the news that our application would have to go before a Committee. Oh yes, and we must advertise our intentions in the local newspaper and The London Gazette. And inform the local Council, so that they could send us *their* set of forms which we would have to fill in - with diagram - before our application could go before *their* Committee. And, by the way, we would have to pay his employers - the local River Authority, for any water taken from the brook. I said something about hoping he wouldn't charge us for cooling our milk can in his brook and something else about building a pond, not a reservoir ... and the pond dream evaporated before our eyes.

That, however, is not the end of the story. While we were still reflecting that we could have gone ahead and had the bulldozer do the job, and no one but our agreeable neighbours would have known, Shirley went to the annual Open Day of a nearby agricultural college. And there, on a trestle table in a tent, she saw a small, free booklet about building fishponds, which told all.

If you fall into the clutches of unsympathetic officialdom, no matter how hopeless the situation, resist the temptation to make sarcastic jokes - as I did. In fact any joking can be dangerous, for life is a serious matter ruled by Forms for All Occasions. Instead, pin on a confident smile - taking care not to overdo it and grin idiotically, as Shirley says I tend to do - and, unless the situation *is* hopeless, show how eager you are to co-operate. And try to change the subject. Beneath every Collar and Tie there was once a heart, and, while stopping short of giving the kiss of life, you might just conceivably set it beating again. Alternatively, agree to do everything asked for - and then do nothing. There is always the chance that your case will get lost somewhere between the In Tray and the Pending File. Time can be your ally. You may appeal. The Collar and Tie might get promoted or pass on, and his successor could have different views. Sometimes if enough time elapses The Law says you may do it anyway - whatever it is - especially if you have already done it.

Shirley and I were tempted to 'do' our pond just the same. But by then the bulldozer had moved on, the rains had come and we were behind with potato harvesting. We still look across that patch of bog though, and if we try a little we can almost see a glint of water broken by the splash of carp, and bounded by reeds and rushes where mallard and heron visit us, and frogs croak amiably on summer evenings.

The family: a survival unit

Success in your venture will depend chiefly on the way you treat your land and the way you treat people. Both will do unto you as you do unto them. Of all the people in your new lives, none will count more than those with whom you venture forth - whether family or friends, you are in it together, and you will have need both to support and inspire each other.

Any group of people facing rural self-reliance should find themselves drawn closer together. If not, there is something fundamentally wrong: wrong with human relationships or wrong with the nature of the venture. If husband and wife are not close together at the beginning they should find themselves drawn closer as time goes by – not in any steady, regular sort of way, for there are bound to be upsets in any healthy relationship, but as each misunderstanding is resolved, the aftermath should be warmer. Not that this way of living is a prescription for ailing marriages. Far from it. I would not recommend any couple or group to embark on it without long discussion in which the best and worst are anticipated and fantasy put in its proper place; nor without a clear understanding of the roles each is likely to play, what is expected and demanded of each one and how each will be affected by the dramatic changes ahead.

Work and play will no longer be mysterious activities performed independently – father disappearing from eight till six, mother confined to the home, children dispatched to school and then subdued by watching TV or being sent out to play. From now on everyone is together. Each has a part to play. The chances are that parents' sex roles will work out fairly traditionally. If mum is the better cook, it makes sense that she should do most of it; if dad is better equipped to hump heavy loads around and wrestle with the rotovator, better he work off his inhibitions that way than by dusting the living room. But if mum is to get outside he should not escape his share of washing up and peeling spuds, nor other jobs which demand neither big muscles nor long experience.

Our children grew up on farms. Although most of the time they were miles from other children and they had hardly any toys, I cannot recall their being bored. When they were not exploring or inventing games they were helping us. The family was a survival unit, they were important, useful members of it and they knew. Today they speak of those days as vivid, happy times. Their lives were firmly rooted in good soil and they have grown soundly in mind and body.

Children fresh from the city will find their lives turned upside down. Unless they are too young to comprehend, they should be brought into the discussions. The family from now on will be working as a *survival unit.*

For a while children will probably miss the excitement of city or suburb – company of large numbers, shops galore and ever-ready TV. In time and with help, however, they should relish instead the open spaces to play in and the secret places to explore, the variety of wild life, the deeper friendship of a few others and the absence of dangerous traffic.

TV watching is bound to be a contentious point. They simply cannot be allowed to spend hours in front of it. Time is a limited resource and if they are to play their part in the survival unit they will have to spend a proportion of their time helping. They can feed the hens, make their own beds, wash up, set the table, shop and run errands - just to name a few useful tasks. At first they are almost sure to resent them, and you will have to be firm and understanding. Remind them of the talks the family had before everyone set out on the venture; remind them that there is no place in it for passengers; that they are important members of the team - equal partners with work to do if they are to enjoy the fruits of success. Make sure they do enjoy them. If you like, reward chores with pocket money. In the days before affluence most children did not receive pocket money automatically - they had to earn it, and this helped develop the sense of responsibility and self-reliance missing nowadays. But see that they enjoy far more than money rewards. Share pleasures with them: bring them into discussions; let them have their own patch of land or some special crops to tend; give them their own livestock to rear - perhaps they could keep a proportion of the profits from any sales they make.

The whole concept of children as a species apart, to be alternately tolerated and indulged, is a modern one. Survival means a return to older values and past customs where, for example, children were treated as young people, protected from harm but bound to do their share of work and shoulder responsibility. This they will only do willingly if they see *the point* of their new way of life, and if they are given abundant but unpossessive love.

Homemade fun

Children will not be the only ones to experience radical changes in life style, however. If you are taking the idea of self-reliance seriously and you read into it more than a blend of escapism and self-indulgence in rural bliss, you will be forced to *make* most of your own fun and satisfy more of your aesthetic demands *close to home* instead of turning on switches and forking out money. The trouble today is that 'fun' and culture have become commodities you buy like any other. In your new life, cut off from the spectrum of coffee bars, 'discos', cinemas, stadia, theatres, concert halls, art galleries and lecture halls, fun has to happen spontaneously and simply - when two or three are gathered together - without travelling long distances and without electronic miracles; while 'culture' has to be raised from its role of spectator sport to a shining quality which pervades daily life.

The change need not be all that difficult. It can be gradual. It is a question of adjustment. Most country districts have their annual fêtes and other social occasions and you can learn to make the best of them. The pub can be a good meeting ground – so long as you can resist the trap of buying unwanted rounds. Friends and neighbours will drop in, and ask you back; and it is up to you gently to lift the conversation from the rut of weather, children and illness when possible. If you play a musical instrument or have a singing voice, that is fine; others around you are sure to, and you can get together; if you cannot play an instrument it may not be too late to learn. And if the *end result* is not quite so perfect as you have been conditioned to expect on radio or TV, nevertheless the *doing* will probably be a lot more enjoyable.

In my occasional bouts of feeling guilty I have agonised over the pros and cons of hi-fi and TV. Shirley and I have a hi-fi, a legacy of our 'better days' but we do not have TV. We have hi-fi because it is there – the set keeps going, it uses negligible quantities of nasty electricity and it saves us large quantities of equally nasty petrol in letting us stay home to enjoy music and plays and so forth. TV we do not have chiefly because I am weak: put me in front of the wretched box and I sit hypnotised through hours of rubbish long after the show I flopped to view has finished. TV simply has too strong a *personality* for either of us – maybe it seems like yet another visitor in our home. And perhaps it is just too much high technology for comfort. Radio is a surprisingly small user of energy and other resources, in manufacture, programme production, transmitting and receiving: (and of course the pictures are so much better than TV!)

Whatever your view on either, one thing stands out clearly: both radio and TV can have valuable roles to play in making country life more attractive, not only for people conditioned by city life, but for those who want to stay put and keep in touch. Radio and TV are not unlike cows and goats – all right in their proper place.

CHAPTER EIGHT

The Livestock

All our animals and birds are our pets; we talk to them, and the really nice thing is the way they talk back to us. Neither of us knows exactly what the other is trying to say, but it helps enormously to cement our relationships. Maybe it is just as well they cannot understand every word, for there are many times when we warn them of a legendary, missionary-size pot we keep handy, and they might take us literally. When any of them misbehave - which is about once a week - we threaten them with this pot or worse. Hortensia, our old, fractious and dearly loved goat, is warned that she would make someone a passable stew. Simba, the cat, is threatened with an untimely end as food for the hens he eyes so disagreeably. When egg output falls I mutter 'Fried chicken' into the henhouse when shutting them up for the night. Only young Buttercup, our clumsy, stupid and dewy-eyed heifer, is spared, even though she would clearly make the tastiest dish of all.

To have or have not

Livestock, it will be remembered, are synonymous with trouble, and so for this and other reasons any decisions about them deserve careful thought. Should you have them at all? Yes, if you eat meat, eggs and cheese. Yes, if you have enough land. Yes, if you can sell your surpluses. Yes, because they have their place in the benign cycle of fertility. Yes, if you love them. But if you resent having to milk them, feed them, clean away their dung, observe them, let them out and shut them up daily, no. If you have an uncontrollable temper, emphatically, no. I have a bit of a temper, Shirley none. I keep mine in check because I know that if you break an animal's trust in you, then you have a long, hard job repairing it, and meanwhile you make life difficult and distressing for yourself. Moreover the animal will probably reward you with less offerings. One way I cope is by removing the cause of situations which, from bitter experience, I know will create trouble. I would say that if ever you come across a worker or helper who maltreats an animal, never let him near any one of yours again.

George Bernard Shaw spoke of 'man's endless slavery to the animal he exploits'. Helen and Scott Nearing – a middle-aged American couple in the thirties – echoed him when they left city life to be forerunners there of today's back-to-the-land movement. In their carefully considered design for living they sought to remain friends with all kinds of animals, and declared: '... we preferred to be free of dependents and dependence. Many a farmer, grown accustomed to his animal-tending chores and to raising feed for animals instead of for himself, could thus find his worktime cut in half.' And they both lived to active old age, legends in their own time.

If you decide against running livestock, life will indeed be simpler for you in many ways. However, you will have a harder job maintaining the fertility of your land; and your place will lack life. Shirley and I began growing a few vegetables almost the summer's day that the property became ours; but, intense though the temptation was, we delayed bringing in our first goats and hens until the following spring. We knew they would make demands on the property and ourselves, and none of us was ready to meet them. When they did arrive the place sprang into life. It became a farm. We had company, and before long our new friends gave us wonderful gifts.

Goats

We had resolved at the outset to keep livestock, but the decision whether to have goats or a cow occupied us for some time. The steep, rough nature of our place and the limited grazeable acreage within its initial 6½ acres all pointed to keeping goats. And yet we heard terrible things about them. John Seymour has said that he knows no one who has both goats and a garden. 'Hardly any fence will contain them and they will ruin your young fruit trees sooner or later – no matter what you do to stop them.' He has declared, 'I like goats one way only – and that is in curry.'

Despite this we opted for goats, and we still have them as well as a garden. To lessen the risks we chose the breed which was said to be least prone to leaping over fences, even though they were not the heaviest milkers. We picked Anglo-Nubians and soon found ourselves in the company of handsome, affectionate, intelligent creatures, who miraculously turn thistles and thorns into milk, and constantly delight us with their astonishing athletic prowess and individuality. Each

From each goat, up to a gallon a day of clean, rich milk

gives us anything from a couple of pints to close on a gallon a day of rich, clean milk which we drink or make into small round cheeses. Now this is not what we were led to believe. Goats' milk, we were told, would taste 'goaty'. And so as a foretaste of our new life, while still in London, I bought a small piece of incredibly expensive cheese from a chic health food shop - and both Shirley and I found it instantly *repellent*. Luckily however the stuff Shirley makes has turned out to be delicious. No two cheeses are ever quite the same, and they are no great trouble to make.

Goat's milk is different from cows' milk - less sweet. But it is not inferior, merely less versatile. The cream from goat's milk takes twice as long as cow's cream to rise to the surface, so if you want to make butter you must use a separator. Now a separator has lots of small parts which must be cleaned after each use, and the nature of goat's milk requires even more of them than for cow's milk. This means a lot of work for a little butter if you are making small quantities. Nevertheless the butter you make will be delicious. It will be white however, though you can colour it if you are not averse to cheating a little.

If you feed your goats the kind of food they like, their milk will taste its best. As with cows, digestive upsets spoil the flavour. Our goats roam freely on all kinds of herbage. Their milk is excellent, as is the cheese Shirley makes from it, and the goats themselves do not smell. Properly kept, nanny goats simply do not smell. Billy goats do, but keeping them is strictly for breeders - and they do not seem to notice it.

I will not pretend that our goats have proved trouble-free. On occasions they have got into the garden and orchard and done some damage, but every time *we* were the culprits, through our neglect or an under-estimation of a bored goat's passion for problem-solving. We soon found that, despite its high cost, pig-wire made the best fencing. We ran it about six inches above ground, putting barbed wire three inches off the ground and plain wire about eight inches above the top to give a total height of about four foot six. We have not yet had a goat jump over or wriggle under it and we find that it also keeps back our Arbor Acre hens, so long as we keep their wings clipped. After much trial and error and stretched tempers we have found that we can keep goats behind an electric fence too. We use four tightly stretched wires nine inches apart, mounted on proper commercial insulators, with the bottom wire about nine inches off the ground. In very dry conditions to make sure, we have run an extra wire connected to the earth terminal and stapled to each fence post about four inches above the ground.

Most goats thrive best on plenty of roughage all year round and unless you are lucky to have ones that do not, you will incur extra work and expense. If you have a bramble-ridden, rough sort of place such as ours was at the beginning, this is no problem; but as you improve it and they eat down thickets, bushes and small trees, you may have to give them hay to replace such vegetation. Now hay is a lot of hard work and not cheap. A cow will milk well on grass alone, but not the average goat. Contrary to legend, she is fastidious and demands nothing less than the best hay all year round. If you give her poor hay she will waste it unless she is underfed, which she should not be. For this reason we decided to limit our goat numbers and run a cow as well and it has worked fine.

Cows

A cow has many advantages over a goat. Goats, when being led have two speeds - full ahead and stop, and you get no warning of change from one speed to the other. Your cow however - especially that most docile and friendly of breeds, the Jersey - will precede you at a gentle pace. She will not find last-minute morsels of irresistible delicacy half-way up a steep bank or just the other side of a fence nearly possible to wriggle through. Neither will she jump over fences, or eat trees - though she may nibble a few leaves quite harmlessly.

A goat's natural time for kidding is in the spring and so you will tend to be without milk after Christmas - unless you have a cow. Although a cow prefers the spring she will breed at any time of the year, provided you can spot that she is 'in season'. Therefore if you keep two cows, with luck you can avoid a dry period by arranging them to calve at different times. Or if you keep one cow and a goat or two, you can have year-round milk by mating your cow to calve in autumn. A cow's milk is versatile. Because her cream has a large fat globule, and so rises readily when set, you can make butter from it without a separator. The skimmed milk which is left will still have some cream in it, and so you can make cottage cheese out of it and then feed the whey to pigs and poultry.

Remember though, your cow will have a huge appetite and she will keep you busy satisfying it. She will need better pasture than goats, with a nice show of clover in it, about a ton and a half of hay a year, plenty of roots or green stuff such as kale, and, if you can afford it - or better still, grow it - a good measure of concentrates - oats, barley and beans, crushed or kibbled. Goats eat a lot too, for they are not the four-legged lawnmowers, tickling their palates with newspapers and pullovers that common belief supposes. But if you have only kept dogs and cats before, the appetite of your cow will astound you, and you would be unkind to her if you failed to anticipate it. To get the most out of either cows or goats you simply must push feed in one end if you are to squeeze plenty of milk out of the other end. When good grass is scarce some of this feed will have to be concentrates such as grains and beans. Once upon a time Britain imported cheap concentrates, but now they cost the earth. For this reason, and on account of the ethic of feeding imported concentrates, you will want to grow your own or else barter. We barter when we can, because we lack arable land, and it works very well.

Not only will a cow or a few goats eat a lot, if you treat them kindly they will also give you an astounding quantity of milk. This should be a blessing, but it can however turn into a problem. Let me recount our own experience. Shirley had to learn how to make cheese, not exactly from scratch but with only slender knowledge to start with. It was a gamble as to whether she would actually *enjoy* doing so in addition to everything else, especially at the end of the day when the time comes for indoor, rather than outdoor, work. Luckily she has found she does enjoy it, accepting that it simply must be done in the morning, even if the garden beckons. The complex, painstaking nature of cheesemaking may not suit everyone however, and it would be a pity if you saddled yourself with more milk than you could drink, only to find yourself dreading cheesemaking - and indeed the somewhat easier job of buttermaking too. Then there is the question of working space and storage. We just about got by when we only had a few surplus pints a day of goats' milk to turn into cheese. Even then we realised that we were stuck with a kitchen too small for cooking, washing clothes, growing seedlings *and* cheesemaking. Buttercup was growing daily, and before long she would be giving three gallons a day. At the flush with goats' milk as well, we could be awash in four gallons a day! It became starkly apparent that for small-scale farmhouse cheesemaking our suburban-size kitchen would be something of a squeeze. To make hard cheese - the kind that keeps - you need somewhere to make the cheese - and that includes some means to heat first the milk and later the curds - and somewhere to store it at a fairly constant 50° Farenheit. We had storage room but not enough for the making of it. Clearly we had fallen into the trap I have been warning you about: to consider all the instalments when you make the down payment on any project - in this case Buttercup. We worked something out, but it was many times harder to find the answer in retrospect.

It is not easy to buy a good cow. I would advise anyone to avoid the cattle market, for you will be buying some dairy farmer's cast-off and there may be something wrong with her. The place to buy a really good cow is the dispersal sale of a high yielding herd where the stock will be guaranteed free from the twin scourges of tuberculosis and brucellosis. You must expect to pay dearly however. It is cruel to turn a herd cow into a lone house cow, dependent on humans for company, and she may give trouble. If you are lucky you may hear of someone who must part with a much loved house cow, the chances are that she will do you proud. If not, and you can wait a couple of years for your milk, do as Shirley and I did and buy a young calf from a good dam.

It is not particularly easy to buy a good goat either. Goats have become popular, prices have soared, and again you run the risk of buying a reject. Your best plan is probably to ask the British Goat Society for the names and addresses of nearby breeders of the breed

We reared our Jersey house-cow-to-be from calfhood

you fancy and approach them. You will soon discover that goat breeders are a special breed in themselves and invariably deeply concerned that their offspring will go to good homes.

However, whatever and wherever you buy, do take along someone, who knows the business, to run his eye and hands over the animal and ask the right questions. We bought our first goat without such help, only to discover that she had had pneumonia at some time and this had left her broken-winded and prematurely unable to carry any more kids.

Rearing males for meat

Remember too that milk is not your only dairy surplus. Half your friends' offspring will be females which you can rear to keep or to sell with no problem, but half of them will be males and the poor little beggars can be something of an embarrassment whatever you do with them. Let us consider this little difficulty first. If you have a Friesian house cow you can rear the bull calves and either sell them for beef or eat them yourselves - that is if you fancy a slice or two of Bill or Buster, or whatever name each has acquired. If you have a Jersey cow mated to a beefy bull, though not before her second calf - again no problem. If she is put to a Jersey bull you can rear an excellent bull calf to eat yourself, but you will get a miserable price if you try to sell it – so low that it will not cover the milk and other feed consumed. Why? Chiefly because Jerseys have a yellow fat which housewives and butchers abhor for no other reason than sheer prejudice. Fastidious shoppers probably suspect they are being slipped horse meat.

It is much the same with goats. You can rear and eat your male kids or sell them – unless, like us you have picked Anglo-Nubians, and these, the Jersey of the goat world, just do not fatten. So whether we are talking cows or goats, what is the alternative if you simply do not fancy eating a member of the family and you cannot profitably sell him? The answer is to kill him, either at birth or as soon as possible afterwards. If you cannot bring yourself to do so, you will either have to ask the vet or an obliging neighbour or butcher. Now I can anticipate the anguished cries of protest from animal lovers everywhere, but the facts are these: if you take an offspring – male or female – away from its mother at birth she soon forgets; but the more time she has to strengthen the bond, the more they will both pine if it is prematurely broken; if you rear a male animal for beef you must castrate it, and this is a mutilation of another living creature which we take for granted, but really creates a pathetic animal, neither male nor female; and finally if you do rear for meat you will feed many gallons of milk which could have been made into cheese or butter, as well as several hundredweight of grain or other concentrates, much of which could either have been eaten by humans directly or fed to more efficient converters of protein, such as poultry. It is often overlooked that beef cattle must eat some *twenty pounds* of protein feedstuffs to produce one pound of the kind of protein meat-eaters prefer. Or to put it all another way: you can either eat your surplus milk as cheese or eat it as beef, veal or kid. And either way, sooner or later, the male offspring must be slaughtered.

If you fancy meat for its flavour and chewability, then that is that. But if you are interested in how much protein, of the kind we humans enjoy and thrive on, each acre of your land will produce, then cheese wins over meat every time, and so you would do better getting your protein from milk. It takes about a gallon of milk to make a pound of cheese, and although this sounds a lot of milk, it represents far less grass and grain than a pound of meat. Even if you threw away the whey that is the by-product of cheese-making, cheese would still be a more efficient way of growing protein. There is no need to do that, however, as we shall presently see.

Running one cow will not cause you much trouble with Officialdom for by making and storing hard cheese and by making butter you will probably consume all of her output which you do not drink as milk. However, if you run two cows or more, unless you can barter in a big way, you will be in the business of selling milk, and then you will need to have your animals accredited. This means not only that calves must be vaccinated against brucellosis, but that all cows and young stock must pass annual tuberculosis and brucellosis tests, and your premises must be of a certain standard. You will be compensated for animals slaughtered for not passing the tests, but the cost of bringing your premises up to standard could be off-putting.

Apart from this, there are restrictions on the sale of cow's milk, butter and cheese. There are no stringent restrictions on the sale of goat's milk, butter and cheese because goats do not suffer from brucellosis or tuberculosis. The Health Authorities will insist that you do not smoke in your kitchen - or wherever you make your cheese - and that you neither wash clothes there nor let dogs in, but this need not pose a serious problem.

You may find that your place already has satisfactory quarters for housing and milking cows and goats. If not, you will have to decide - as we did - where to site them. Living in a very old cottage and farming very steep land we knew we would have to expend a great deal of energy - climbing and descending steps and steep slopes, often carrying loads or leading livestock. And so when the time came to decide where to house our goats and eventual house cow we resolved to have them close to the house. We knew the number of

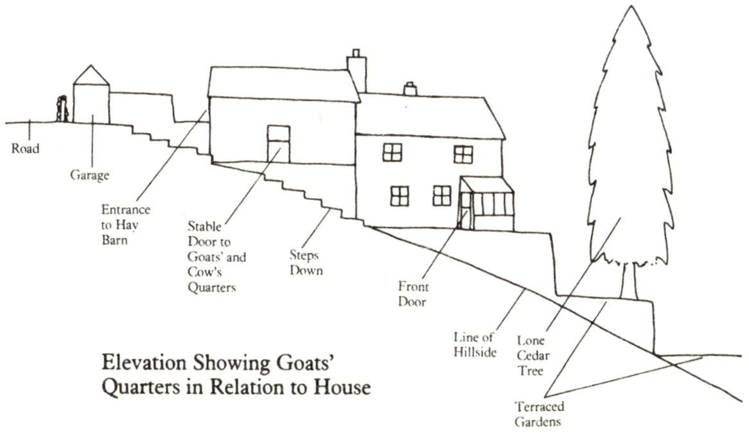

Elevation Showing Goats' Quarters in Relation to House

journeys each day that stock demand for milking, feeding, cleaning out and routine inspection, and we wanted to minimise the effort. At first we thought conventionally of a brand new shed just along the track on one of our rare pieces of flat land and level with the kitchen. Goats however are best insulated from excesses of heat and cold and they will not thrive in draughts. Clearly a new shed would mean either thick walls or insulation, and careful ventilation, in fact a major building job. Surely, we thought, there must be a simpler answer - and there was: to use existing space in the two-storey building above and behind the cottage - the spot we came to call 'the barn'. Its walls were over two feet thick, we had already repaired the roof, and it could not be closer - right at the top of the spiral staircase from living room to Shirley's workroom! To play safe, we checked with the Health Authorities before proceeding, and received the 'all clear'. It works handsomely. We have no problem with flies or smells and by siting the milking parlour right next door to the animals' sleeping quarters we have ensured that there is the least possible walking distance for both humans and animals. The arrangements might not pass the test if we were selling our milk products, but since we merely barter any surplus we face no problem there.

Sheep

Sheep are not much bother to keep. On good pasture you can run three ewes and their lambs to the acre. The ewes will live on grass, their lambs will fatten on it, and you will rarely have to feed hay.

If you are part of a meat-eating community, sheep may suit you fine; if not there is the problem of the ram. You will find it uneconomical to keep a ram to serve less than about two dozen ewes. Now at most a family will need no more than the lambs from half a dozen ewes, so you must aim to share a ram with three other families or else run a couple of dozen ewes and a ram - if you have the land - and sell the surplus lambs.

Your only problems are likely to be fences, worms and flies. To keep sheep where they should be, sheep netting is essential - and it is not cheap. To protect against sheep fly - a green blow fly which infests sheep with maggots - you must dip the sheep or spray them, unless you are safely above the 700 feet altitude line. You might have trouble with ked or sheep-louse, but the worst enemy of all is worm infestation, a problem resulting from over-stocking. To avoid it you rotate your sheep around the farm, always following the cows - never before them.

Pigs and poultry

Pigs and poultry on a small scale can fit neatly into a self-supporter's economic plan. Poultry will give you free eggs if you run them free range so that they get as much nourishment as possible from grubs, insects and grass seeds and dropped cereal grains. Like pigs they will still want your household scraps and any grain you are able to grow. And if you cannot thresh it, they will pick out the best and leave you straw you can use as bedding for your cows and goats. You may still have to throw them a certain amount of bought-in feed, but the art of the game is to keep this costly proportion of their diet as low as possible.

We run enough hens to keep us in eggs all year round and give us a surplus which we sell and barter so that with luck they pay for the wheat and occasional layers' mash we buy. The number varies between ten and twenty depending on age. Nigel, a bantam cockerel, keeps them happy and generally organises the hen run. They eat up all our scraps and small and spoiled potatoes, and they are on free range - the *only* way to keep hens. They graze busily as a flock and they are a delight to watch. They keep healthy. I made their houses almost entirely from waste material, chiefly wood, and they are quite handsome structures.

We should of course buy day-olds and rear them, but because we are always trying to save time and keep life simple we have so far bought at point-of-lay - about 20 weeks old. We picked Arbor Acres from the outset and have kept to them. They are friendly, not a bit 'flighty', rarely go broody, graze heartily - even in the rain - and lay beautiful, big, brown eggs. You have not tasted eggs until you have eaten your own!

Chiefly because we eat so little meat we have no pigs, but they are certainly worth considering in any livestock plans. They will gobble up your milk surpluses, cultivate your land, root out bracken and other weeds, and give you good meat. However, as I said, unless you can grow most of their food, and you can keep them behind an electric fence when folding them - and not get too fond of each one - like *all* livestock they will cause their share of trouble.

We would not recommend planning to run large numbers of pigs or poultry. Better to keep just enough for your own needs, to mop up waste and surpluses and pay for their bought-in feed through their own surpluses. If you expand and run them more for sale than self-sufficiency you run the risk of 'buying retail and selling wholesale', for you are the merchant's customer, and unless you can sell at the farm gate, the retailer's supplier.

All these points are well worth long discussion, for they are crucial to both your economic plan and your long-term plan. Shirley and I planned to rear our female kids and calves as replacements when necessary and to have unwanted males swiftly killed. Pigs did not fit into our plans, so we planned to feed our poultry as much whey as they would drink and barter the rest with a pig-keeping neighbour in exchange for the odd bit of pig meat.

As you can see there is a great deal to keeping a goat, a cow - or *any* creature for that matter, including the honey bee - and unless you are prepared to learn about its habits and needs beforehand, as well as later, then you would do better to stick to fruit and vegetables. They too may feel pain, but they do not cry out or look at you reproachfully.

CHAPTER NINE

The Wherewithal

Most books and articles about living on a little land avoid the subject of money. People who have made enough money to buy land often feel guilty about having done so, especially when confronted by eager, young, would-be self-supporters with the embarrassing handicap of being penniless. The subject can be acrimonious. Also it can lead on to politics, and since politics - like religion - makes most people uncomfortable when confronted by anyone stupid or selfish enough to hold an opposite viewpoint, it too is avoided in favour of safe subjects such as ley lines, windmills and food.

All the same we simply must grasp this nettle, for whether we like it or not, money plays a key role in turning the dream of country living into reality. As things are, anyone can own a piece of land however unfitted he may be. He can sit on it and let it go derelict or, worse still, attack it with machines and chemicals and ruin it. Just on account of having money. Meanwhile, countless thousands of good people - young, strong and bright - are itching to get out of unhealthy cities and grow things on land, build their own houses, become self-reliant and stop being a liability on the environment. But they cannot - just on account of having no money.

Shirley and I were among the lucky ones. Neither of us had caught the cigarette habit, nor had we spent at all lavishly on drink and entertainment; and so we had managed to save over the years. By throwing myself into my work with misplaced and unappreciated passion I had steadily pushed up my earnings too. But our chief good fortune was the property boom of the Sixties. We had bought our apartment before prices soared, and so when we came to sell we had enough to exchange it for our run-down place. To revive the place, we used our savings, both as cash and cashed life assurance policies, a small legacy and my redundancy pay. These together yielded just enough money for converting the house, developing the land, a modest venture into alternative technology and some working capital.

Our good fortune is the kind that is more likely to come to you when you are not-so-young. When we were young we bought our first farm quite differently – with a big mortgage. But then we had 200 acres and 33 milking cows producing the kind of income that allowed us to pay off the mortgage and live. Then again 'luck' played a major role – so big a role in fact that Shirley and I no longer believe in pure luck, but in The Flow. You have to use what you have. And if you lack money you must rely on other resources – of ingenuity, patience and zeal.

Many families, couples and individuals *are* managing to buy places and they are doing so by getting together, mustering all their resources, pooling them and buying small acreages with big houses, to live more or less comunally. At the time of writing an average contribution of around £5,000 can make such a scheme work, and if some people put in more than the average and some less, it means that those with £2,000 to £3,000 do have a chance. This may sound a lot of money, but if two people work hard for a year or more while living more simply on one person's income – as some we know are doing – and then sell a car and anything else not strictly needed, and add this to whatever savings they may have started with, the sum raised can be higher than expected in less time than you would think.

Some self-supporters we know got their capital together by buying old cottages with the help of big mortgages, doing them up and re-selling them. With luck, judgement and a lot of hard work three such moves and transactions can transform an initial £2,000 or so into £15,000. What is more, they learned all they needed to know about building and quite a lot about gardening on the way.

Another less risky way of acquiring a smaller sum is to take a job on a farm for a while. If you are without experience you may have to work for board and keep at first, but before long you should be worth a full wage. If you are lucky enough to be able to run a house cow and work a fair sized garden you should be able to live very cheaply, save and – most important – learn as you go, not just about farming but about living in the country. If you find you have to swallow some organic principles it will be a small price to pay for what you may learn. And who knows? Someone in the district may be able to help you get started – the country is brimfull of kindness and The Flow may work for you.

While I would not recommend 'squatting' as a solid foundation for building your future, it is possible to find unused land with no apparent owner, and it may be helpful for you to know that an owner loses his right to reclaim possession after 12 years. A squatter can then claim what is technically known as a possessory title. For some years two friends of ours have been quietly cultivating an unused field next to their cottage. If the owner turns up they will have gained a lot of food; if he doesn't it will one day be theirs.

I would strongly urge you not to start with a mortgage or other large, long-term debts. If you do, life will be like trying to fill a cider barrel with the bung missing. By all means have a go, if all you are committed to is a smallish, short-term debt, quickly liquidated from the proceeds of a well-paying craft or some other source of income which you may have up your sleeve. But *please* do not saddle yourself with the debt that never grows less; far better to wait a while, build up some capital, pool with others or - however regretfully - postpone the idea of living on a little land, and devote your energies instead to some other 'craft' which demands less initial outlay. The change in plan need not be permanent. Remember, there is always The Flow; and if you keep your dream alive by meeting people with similar views of society, while remaining vigilant for opportunities, something will probably turn up.

Several of the helpers who have stayed with us for a week to a month or more have been 'professional wwoofers', travelling from farm to farm and acquiring friends and skills as they go. This can be one way of keeping the dream alive. There are others. In the spring of 1977 two conservationists, David Stephens and Paul Holder, launched a non-profit making company, Ecological Life Style Ltd., specifically to enable people to settle on the land without having to buy expensive properties outright. Prospective settlers were invited to subscribe money in monthly instalments, which would be converted to Ecological Land Bonds, index-linked to the cost of living. As funds grew the company planned to buy properties, situated as near as possible to areas preferred by the greatest number of subscribers. The company would encourage subscribers to group together and form co-operatives, working full time, or part-time at first, and eventually living in. Each co-operative would lease its land from the company. (Full information will be given if you write to the address in the Source Guide.)

There is nothing sacred about the status quo. At any time the overdue reaction against monoculture and factory farming may get under way, the people may take note that although large farms produce more per worker, they produce less *per acre* than small farms. They may note also that the limited resource is not workers – as unemployment figures repeatedly show – the limited resource is *land*. Industry is failing to attract the recruits it demands; cities are becoming harder to finance and administer, and they are demonstrably less attractive places to bring up families. In this climate radical movements for land tenure reform, such as Land For the People, The Diggers and The Rural Resettlement Group (see Source Guide), are becoming increasingly active, and the day of new thinking on agriculture and accompanying reform in land tenure may be closer than we expect. When it comes, and the countryside receives the injection of brains, brawn and money which it so richly deserves and has so long been denied, the smallholdings, which may be created from large farms and estates with low output per acre, will be run by people with demonstrable qualifications. If you can show that you have worked on the land and that you have other supporting skills into the bargain, you will be in a stronger position than other claimants. It is a scenario which may come about, or it may not. But it is one to work for, and the work itself can be rewarding.

Earning an income

Once you have scraped together the indispensible capital to get started, money still comes into the picture, painted large. You will still require an income. If you have enough land, or you are intensive, surpluses which are left after you have grown enough for your own appetites may add up to a worthwhile income – eggs, honey, cheese, butter, fruit, vegetables and possibly pigs. I know one self-supporter who is planning to breed house cows, though I worry that if he breeds too many at once they may miss their sisters too much when sold.

Of course if you can specialise a bit and grow worthwhile cash crops, the land itself will go a very long way towards generating what you need. It might even provide all – potatoes, tomatoes, lettuce, soft fruit, asparagus, peas – the choice obviously depends on your aptitude, your land, climate and, of course, markets. It is simply no good at all having a fantastic outlet ten miles away if the cost and time of transporting your produce erodes your profit. And it is not much better having one close by – possibly a local shop if, when you tell them you are about to pick, they inform you they are sorry but their 'normal' supplier has been delivering for the past week. You must nail them down. And you must deliver good produce on time, in the required quantities, of the right size and quality at the right price. All this is something to discuss and tie up securely well in advance.

Your closest outlet is, of course, the farm gate, but even here you should think a little. Are you on a sufficiently busy road? Are its users likely to buy, or do they mostly grow their own? Is your house close to the road? Your customers will not relish a 400 yard trek up and down a steep muddy track to buy one lettuce. Can you dispatch them reasonably swiftly, or will you feel obliged to spend 20 minutes chatting before and after you have sold the lettuce? While time is one of your constraints, this problem cannot be evaded and you may have to develop one or two techniques 'to speed the parting guest'. And suppose you are away - will you mind strangers roaming around in your absence?

To sell some crops profitably you certainly want to catch the market; some crops you will probably have to grow under glass - either in cloches, Dutch lights or a greenhouse - and that demands capital outlay. Almost everybody seems to be switching from glass to polythene chiefly because the outlay is lower and labour costs less. I would not recommend doing so; glass, if you are careful, lasts forever, whereas plastic gets brittle in sunlight and lasts only a few years. It is part of the 'throw-away' society, and since it is made from oil, users are in for a shock on price when they come to renew it; and the whole concept is 'unecological' anyway. You can take advantage of this trend. Shirley and I have bought 100 cloches and some Dutch lights in near-perfect condition for a fraction of their true cost.

How much income will you need?

Chiefly this depends on how many of you there are. Shirley and I are lucky in a way for our children were grown up and living their own, separate lives when we made the move. On the other hand we feel the effects of two of the three basic constraints - time and stamina - more than younger people. Now that the full flush of the first years is over, even though we are probably twice as fit as others of our age, we have to admit that much of the outdoor works takes longer than it once did. And so on occasions we must *buy* time, by paying others to do what we can no longer do so well - and this in turn increases the need for income from writing or other sources.

Inflation is a wrecker of plans and budgets. To fight it we strive to reduce our income requirements by living even more simply than before. We live by five principles. When the time comes to buy anything we ask ourselves 'Could we do without?' and even now, it is surprising how often the answer is 'Yes' - not all at once perhaps, but over time. Once we drank at least six cups of tea a day; now we have rationed ourselves to two. If we cannot do without something, we ask ourselves whether we could make whatever it is ourselves - from coffee beans, to glasses or a table. (We are drinking dandelion coffee - we actually like it; we have found how to make glasses from old bottles; and I have a simple design for a table - all we need now is *the time*!) Then we see whether instead we could share with someone else; we have already done so with farm machinery and we are extending the idea to household things; eventually it may include sharing vehicles - already the most costly item of all. Fourthly, if we *must* buy we try to buy secondhand; from sales, jumble sales, Oxfam shops, secondhand shops, deals with neighbours - all are lucrative sources. And finally as I have already described, we avoid wasting *anything*. In short, the five principles are: do without, make your own, share, buy secondhand, and avoid waste. In this way and others we are able to keep our income needs as low as possible - and, because the principles fit our philosophy, there is little strain in following them. In this voyage of discovery which is part of learning to live on a little land, we are beginning to distinguish our needs, which are small, from our wants which can be insatiable. And as we do, we are not finding ourselves deprived at all.

Money-making ideas

If as is likely, you cannot generate all your income from selling surpluses and growing cash crops, you have three other choices: you can work at home on some craft; you can run your own business which may involve *selling* your skills; or you can go out to work. I suppose that at a pinch you could undertake all three, but if you did I suspect your weeds would soon be higher than your crops!

Let us look at activities with more of a craft 'feel' to them first, while remembering that all of them produce goods or services for people, and so firstly must fulfil a need and secondly must be *sold* - distasteful though the use of business acumen may seem to some of us.

In just about any district that I know anything about, there is a chronic shortage of more or less skilled people willing to tackle building, carpentry and odd jobs. This whole range probably presents the most golden opportunity of all. I have a number of friends who have picked up their building knowledge in various ways - building or converting their own houses, working on building sites and so on -

who are now in demand for work ranging from house conversions to putting up shelves. Apart from general building, useful skills are painting and decorating, stonemasonry, bricklaying, glazing, plumbing, electrical writing and carpentry. Allied with carpentry are cabinet-making, furniture restoring and interior design. Wood is perhaps the most satisfying medium of all to work with and once you are skilled you can happily and profitably turn your hand to making beehives, gates, rustic furniture, cupboards, doors, picture frames, panelling and toys. Again, allied with furniture-restoring there is antique dealing, for once you begin picking up old furniture at farm sales, auction rooms and less likely places, you will come across chests of drawers, cupboards, tables and chairs, probably coated with horrible paint and in various stages of decay, all of which are a delight to save from the fire and to turn into lovely furniture.

Wood-carving and turning are fascinating and satisfying crafts but it must be admitted that the spoons, lamp-bases and bowls you turn out may not offer you a very attractive hourly rate. (Though if you start doing such sums almost *any* craft begins to sound dubious!) As with pottery, so much depends on markets. If you happen to be in or near a tourist catchment area you stand a better chance - always provided you are willing to bend to popular taste and not insist on pure aesthetics. Be more than a little cautious about pottery. It is the most capital and energy intensive of all the crafts, and your apprenticeship will be long. You may thoroughly enjoy 'one-off' designs, but when you have thrown your umpteenth saucer for a lucrative restaurant order, your enthusiasm may wane.

One of our best friends seems to make a handy income from leathercraft, which he had the foresight to learn from a skilled craftsman while he was planning his new life in the confines of his city flat. Over three years he has built up a lucrative small business specialising mostly in hand-carved and intricately tooled belts. Jewellery-making is another labour-intensive craft which can pay, though once again, ready access to one or more good outlets is essential. At least one person near here makes some useful money at signwriting - the sort of craft that grows through recommendation, for every good job done is an advertisement for the next one.

Repairing things strikes me as mid-way between a craft and a trade. Good repairers are always in demand. If you can fix vehicles, tractors, other farm machinery, garden mowers and the like, as well as household appliances, radio and television sets, you will be one of the most sought after people in the district.

Sedley Sweeney, self-supporter in Wales, was able to buy a forge from the smithy in a nearby village when the blacksmith retired. With it he is able to do most of his own implement repairs and make new tools and equipment to his own design. He can also generate income by doing work for others, and by making items for sale such as cheese presses and wrought iron work. If you are handy at metalworking and you can get hold of a forge, a welding kit, a lathe or any other useful equipment secondhand, you can be virtually certain of a worthwhile income. Our society is on the brink of emerging from the 'throw away' era to one where the high price of materials, all growing ever scarcer, will force us to re-use and recycle more. If you are the sort of person who can improvise from scrap – in wood, metal and plastic too – now is the time to seize the opportunity.

Despite anything I have suggested to the contrary, writing is a means of livelihood which is well suited to the life of a self-supporter. I make a small income from it, and I could probably make a lot more if I was prepared to write about sex, crime and corgis. However most of my books seem to touch too many exposed nerves, forcing the reader to think about the unthinkable and leaving him feeling uneasy and even guilty about the unecological way he and most others live. (Though *this* book may be the beginning of a breakthrough for me ...) I write articles too and do the odd talk or interview for radio and TV. I make a little more than if I gardened by the hour on someone else's place; but I'm not sure if I would be happy about doing so when there is always so much still to be done on my own place. Apart from that, I feel that if I have any writing ability at all, I should use it to hasten the gentle revolution, however small the contribution may be.

Almost anyone can write. All you need is something to write about, elementary grammar, humility, sensitivity, good powers of observation – and persistence. You need the last quality in abundance, not only because the only way to learn to write is *to write*, but the only way to get it published is to keep trying until you have established some successes. You also need the first qualification – something to write about – and do not imagine you lack one; that you are reading *this* book sets you apart a little. If what you communicate is what people would like to hear, then not only will you earn a few pennies, but you will have gained a lot of satisfaction.

What else? Moving away from what might broadly be called 'the crafts' we encounter a whole range of services you could offer if you have the skills or can learn them, and if enough people want such services. We know of one lass who used to work for a book publisher as an editor and now does the same work on a free-lance basis. Several others are working as indexers. Some do home typing. We know of one or two people who give music lessons - guitar, piano, violin - both to individuals and to groups. They all enjoy the contacts this creates. We know of two young men who, quite independently, are continuing their old work as computer programmers. They keep in touch with their clients by telephone, letter and occasional trips to the city, and they make more than enough to live on - enough to save for the future or to plough into their land. And all without the hassle of commuting and the constraint of keeping regular hours.

The invention of the small offset litho printing press has revolutionised printing, making it possible to produce short and long runs of high quality work in one-room premises. If, alone or with one or more others, you can add design to printing know-how, you may be able to develop a modestly profitable small business printing anything from circulars and letterheads to booklets and periodicals - whether the parish magazine or an 'underground' newspaper.

One enterprising young man, who visited us recently, has a van which earns him money by moving goods and people. The occasional advertisement in the press, a few cards in shop windows, as well as recommendations, keep him and his van busy. You can consider using your farm vehicle for outside work, but you should have a word with your insurance company beforehand. There is no reason why your tractor and other farm machinery should not earn you money. When I was farming before, I set up quite a seasonal business mowing hay and making silage with a buckrake on my small tractor. For contract work rotavator, mower, hedge-trimmer and chain saw are the machines which spring most readily to mind.

Provided that you really do have the right skills, you should be able to offer useful services to local farmers. They are always glad to know of reliable relief milkers, and anyone able to do a good job of ditching, hedging, and fencing and - in some districts - dry stone walling too. You will make some good friends, if you can deliver the goods.

We know of several people running successful stalls in their nearby market towns. They are selling their own produce, backed by grains, dried fruit and so on, bought in bulk and repacked, as well as a variety of crafts. A stall or shop is essentially an enterprise for a group, such as a commune or a co-operative of growers.

As a development from sales at the farm gate you might consider catering. You could try the temperature of the water with morning coffee and afternoon teas, and if it appeals to you and the customers come, expand it to lunches as well. If you are in a scenic district, once the message spreads that you are serving wholefood, mostly home-grown, you may expect a steady stream of walkers, cyclists and motorists. Bed and breakfast is another development that a number of self-supporters have undertaken. It is best to build up 'regulars' you like, and who are sympathetic to your ideas.

You could look into the idea of allowing campers and caravanners on your land. The venture can be lucrative. Although planning permission is normally not required so long as you don't exceed 28 days a year, it would be well to ask a friend to enquire for you through the local planning office before proceeding. Nothing is simple with bureaucracy, especially if you want to exceed the 28-day period, for you might be asked to have a site licence.

When you put your mind to it, the opportunities for making things and providing services are almost limitless. Have a 'brainstorming' session with your family and friends and write down all the ideas – however silly – that you can think of. Then go through the list and pick out any that look like starters and explore the market. Here are a few more ideas in addition to those I have mentioned: babysitting, crocheting, knitting by hand or machine, mushroom-growing, photography, plant propagation, poultry breeding, scrap metal dealing, weaving.

If you find that you and your friends together can offer a useful range of services, you might consider setting up a central agency with a name, address and a phone number for prospective customers to contact. It would require a little organisation, but it could help to bring in the business and share it around equably.

One word of warning may not go amiss. If you intend to carry out any kind of business on your own place, check with the Local Authority first, either directly or through an intermediary. Although you could probably sell from the farm gate, 'do teas' and offer bed and breakfast without ruffling officialdom, a fully-fledged restaurant or a thriving printing press might offend some by-laws and bring the bureaucrats to your door in howling packs.

Unless you have a skill which does not depend on local markets or swift access to a particular city, and unless you are able to 'cheat' and enjoy unearned income, you must depend on what the district offers as reward for your efforts. If you are able to work from home, you are among the fortunate; if not you will have to go forth and bring in money from outside, possibly by working for others on a part-time, seasonal or full-time basis. The prospect for work, therefore, may be something to explore when you first consider a district with a view to settling there. Do not just assume that work grows on trees.

As Robin Clarke in Shropshire emphasises: 'Don't make the move to the country thinking you can pick up a three or four-day week job just like that. You won't be able to. Some jobs you can rely on. A few weeks work a year at a nearby farm during the hay, corn and potato harvest is a pretty safe bet - but it won't be enough.' He points out that teaching is the best job to go with this life style, but such jobs tend to be scarce for that very reason. He stresses the importance of ensuring an income of about half what you were earning in the city. 'With that you've got a good start on the road to independence,' he says. 'Without it you'll end up working a 5½ day week at the nearest Chuckie chicken factory.'

One or two of our friends take full-time gardening jobs for a while, either when seasonal work generates a demand or when they need the money. I have sometimes thought that if I had to look for work I might seek an acceptable company for which I could be a local sales agent. They might even let me operate independently, rather than put me on their pay roll. I have known resilient characters work in quarry, sawmill, factory and shop. Some with well paid skills, acquired in their 'better days', are able to practise them on a part-time basis: architects, systems analysts, commercial artists, nurses, engineers ... One couple with no livestock to tie them, work for weeks at a time as chauffeur/handyman and cook/housekeeper together in more or less stately homes.

Whatever you do to bring in the money, try not to let it compromise your ideals; do not fall into the trap of becoming over-dependent on remote people who cannot be expected to care about you; and do not lose your dearly-won freedom.

A degree of insecurity will always accompany you in your new life if you are living it as you should. Insecurity is part of the process of developing self-reliance, and you must learn to live with it. There is no need to fear. Provided that you do all you can in the life, you may depend upon The Flow to take care of you.

Finally, and above all else, let not other work unduly interfere with what you set out to do in the beginning, for your land must come first. If you ignore the land, you do so at your peril. Which is surely a maxim the whole world might heed.

CHAPTER TEN

The Chattels

'One of my biggest criticisms of the self-sufficiency movement,' says Robin Clarke 'Is that it is so food centred - as though food were the only thing we ought to be trying to raise ourselves. It isn't'. And he reminds us that we all need clothes, homes, materials and energy as well. When he left his well-paid desk job and centrally heated London house to build and farm, by far the most important discoveries he made were his own unsuspected abilities. Within 18 months he changed from ham-fisted handyman to a passable tradesman, familiar with concrete-mixing, drain-laying, carpentry, joinery, roofing, plumbing, wiring, guttering, rendering, farming and vehicle maintenance. And able to do many of them well.

I share Robin's views - and to a lesser extent his aptitude. I have found that as I break through successive barriers of mystique about each craft I emerge more confident and able than I would have dared to hope. I realise now that for most of us city life encourages helplessness. It is all too easy to pick up the phone or nip round the corner for help; and help arrives, in time but at a price. In the country things are different. For a start the nearest help is likely to be a long way off. If a tradesman has to travel to you he will charge his time and travel costs. If you contact a neighbour he will probably be busy and understandably expect you to bide your time. He may put right whatever is wrong the first time he comes, but he will be justified in expecting you to fix it yourself the next time round. And so you have to nourish your self-confidence, read all about it, watch others at work and then 'have a go'.

One fact stands out clearly. The more chattels you have - be they household appliances, vehicles or farm machinery and equipment -

the more there is to insure, store, break, break down, rust, wear out, mend, maintain lose and replace. My advice, born of character-building experience, is to follow the five principles I have advocated in the previous chapter: do without, make your own, share, buy secondhand and avoid waste. They apply to *all* chattles, in the home and on the land. Let us see how these principles can affect the way you acquire and use farm machinery and equipment.

We do not believe that it necessarily makes sense to do without farm machinery, but you can aim to keep it minimal. You may remember that Shirley and I run our place with just a little old van, a rotovator/trailer, half an Allenscythe mower and half a chain saw. If we encounter any job too tough for these machines to handle, we call in a neighbour or two with machinery big enough , working either on a straightforward contract basis or by bartering. This way, much of our heavy development work has been done by machinery, but without our having to incur a heavy initial investment in it, and without too much of the cost and hassle of maintaining it. Inevitably there have been frustrating delays when a neighbour has failed to turn up with his machine on the day appointed, usually for a very good reason; but at least our place is not littered with great hulks of machinery, mocking us in idleness for months on end, and in a small way we help to keep neighbours' machinery on the go. Our place is not typical, but you would be well advised to heed our experience if

Seeding with a borrowed 'fiddle' helped minimise machinery

your acreage is small, and if – like ours – your fields are odd-shaped, tiny, steep or not easily accessible.

For a place such as ours or smaller, a rotovator will probably be enough, though do make sure you choose the most appropriate model. Fundamentally there are two breeds of rotovator: one has powered wheels; the other relies on its tynes to propel it. The first breed is more expensive but, because it can cultivate more at a time it is better for larger acreages, it is also better for hilly land because the wheels aid traction and provide more control. It is the kind we chose and on the whole its 5½ horse-power does a good job, but, like a lot of farm machinery, its performance still leaves plenty to be desired. I think engineers design things for ideal conditions. I have this image of them trying out their pet prototypes in imported loam on an impeccably level office floor and drooling ecstatically at the way they perform. Real life can be different. Even before I added the trailer hitching attachment, I found that my model rotovator was top heavy. And as if that wasn't enough I found it lop sided too, because the engine and fuel tank were mounted off-centre. Time and again it has turned over on our slopes, and it weighs a full 350 lbs: now I have learned that I can minimise this risk by always turning so that the heavy side stays on the uphill side, though it is not necessarily convenient to do so. If I try to rotovate uphill, the wheels often slip and dig a hole from which the only way out is to reverse. So I have learned to drive the thing uphill with tynes raised and then turn, stand precariously on the splashguard at the back, and literally *ride* it downhill with the tynes working. Unless I put all my weight on it to force the tynes to penetrate the ground, it is liable to get an idea into its head that it is Concorde and try to take off – all because traction is transferred from the wheels to the tynes which rotate umpteen time faster! I make this point at some length because some advertisements for rotovators show well developed young ladies in tight jeans working the brutes with one hand. Shirley who can hump heavy sacks and hack brambles and bracken happily, sensibly keeps well clear of it. As with so much about living on a little land, real life simply does not correspond with the dream – as my aching bones remind me half the night after only a few hours of hilly working.

For the small farm and the larger garden, a tractor and some implements will represent a worthwhile investment. Pick a good smallish secondhand tractor from a reliable source. Basic attachments for it are likely to be: trailer, link box, plough, disc harrows, spike-tooth harrows, chain-link harrows and potato ridger or furrower.

Buy all these secondhand, except perhaps for the trailer, which you ought to be able to make yourself, if you can find the time and someone to give you a hand. If you can share any of the equipment, so much the better. Disc harrows help the sod to break down quickly; if you cannot pick up a set cheaply, you could manage without by allowing more time for natural breakdown and by harrowing more often with the spike-tooth.

Think twice about running a car, van or pick-up. If your place is near a bus stop and your scale of operations is so small that any surplus can be sold at the farm gate and through local shops or hotels, you might be able to manage without running a vehicle at all - especially if you could share. This would save you your biggest single outgoing - anything from £200 to £750 a year. If not, you will want to manage with the smallest and most versatile vehicle you can find.

Because buses pass the door and we are close to the village, Shirley and I could run our place without a vehicle by sharing transport if we were really pushed. Until then we shall continue to run our little van. Soon after we arrived, when we no longer needed our original Dormobile van, we sold it in favour of an imported Citroen 2CV van. It promised to be ideal - the closest thing to an 'alternative technology' vehicle one could wish for: no fancy chrome or unnecessary gadgets, up to 50 mpg, 5cwt. capacity and an incredibly smooth ride. We sold it after a year because what we gained on the 50 mpg 'swings' we more than lost on the spares and maintenance 'roundabouts'. And so we settled for something still small but more conventional - something that would neither make young Tom wince when I asked for his skilled help as a mechanic, nor make me frown when I got the bill for imported spares. Reluctantly we exchanged it for a cannibalised Morris 1000 van guaranteed by Tom to run and keep running. And it has - touch wood.

You might find a pick-up even more versatile than a van; however, secondhand pick-ups are in great demand and you may have difficulty finding a good one. If so, look for a good van. Do not hang on to your city car - you will find it neither convenient for taking a randy goat to the billy nor ideal for humping rotting manure around. You will not often be making long journeys and your van or pick-up will quickly convert to a luxurious limousine with the help of a stiff brush, a piece of old carpet and a few cushions.

Machinery does save time of course and so it is a powerful means of releasing you from this most pressing and basic constraint. However, unless you are careful the saving can be largely illusory. Remember that unless you were one of the lucky ones, you spent time in the first place earning the money to buy the machinery; you may have devoted time to building a place to store it; you will certainly spend time repairing and maintaining it; and when the day comes that it breaks down on the job and leaves you stranded, you will curse the wasted time that results.

All the same, I do not suggest you do without farm machinery and equipment altogether, unless your place is small and easy to work, and unless you believe so fervently that industrialisation is an evil to be avoided at all costs in every way that you feel obliged to set an example. I do urge you however to think carefully before acquiring any piece of machinery or equipment, and consider instead the alternatives of making do and sharing.

Improvisation

A little land can give you enormous scope for improvisation. Survival brings out the inventor in you and releases floods of creative energy. The ingenious way the trailer fits onto our rotovator is one piece of improvisation that I bless. We have many others: helped by a friend I harrowed all our pasture sowings with a set of homemade

Harrowing a pasture seedbed with a brushwood improvisation

brushwood harrows in the old tradition; I made our henhouse from scrap wood so that it exactly fitted an old bed base; through two successive drought summers we watered by gravity, simply poking the end of a considerable length of half-inch hose through a tiny clay dam in the course of our spring; we had a friend build us a sturdy milking stand for our goats, again entirely from scrap wood. Now all this saved us having to buy equipment, whether new or secondhand. And it gave us a wealth of satisfaction in the doing.

There is a snag to improvisation however, which I feel bound to mention. So long as money is scarce and time is breathing down your neck, you will tend to make do with gates and shelters hastily knocked together from scrap, indeed with all manner of things made the same way. They will constantly offend your eye, they will not last, and one day they will probably collapse and have to be made again. However, there is *another* way of doing things: beautifully - the way the Shaker Communities of America during the last century and before made *everything*: lovingly. Then all around them was an expression of their philosophy, and a testimony that they *cared* about materials. The Shakers had no time for art as something apart from life, and so they put all their artistry into functional things. Everything was a joy to make, to see and to use. And as a rule anything they made only had to be made once. I have seen their works, and the Shaker approach is an ideal I cherish but seldom achieve. It would be so satisfying to be able to spend a week making a really handsome gate, sawing the wood from the whole tree and fashioning the metalwork myself, but there simply is not time. And so regretfully we buy new or secondhand, or knock up something in a hurry that looks like what it is, and will not last. You will probably be forced to do the same, which is a pity. For all along you will be reminded - as we are - that there is a certain rate at which things disintegrate, rot, rust and wear away; and there is a certain rate at which you can make things to counter this process. In this you are constrained, as I have pointed out, by time, money and stamina. The art of running your place, therefore, is not confined to growing crops, husbanding livestock and generally organising: it is your skill in ensuring that the rate of disintegration - or entropy, is less than the rate at which you can prevent it. And if you learn nothing else from the reading of this book, the reading will not have been in vain!

Sharing

To 'roll your own', to do your own repairs and to improvise fully, you will want a well equipped workshop. You will want basic carpentry, building, and engineering tools, and these should preferably be bought secondhand; though, if not, they should be good quality and neither the DIY gadget breed nor surplus store 'seconds'. And you will deserve a warm, well-lighted room, for to work on the kitchen table or in a corner of the sitting room is to invite divorce or worse. The heart of your workroom will be the work bench, and along the back of it or above it you should have a tool rack. Sooner or later you will want a pair of saw-horses, so you might as well make them for a start – the practice will warm you up.

Throughout this book I have extolled the virtues of sharing; and in the matter of machinery, equipment and tools there is ample scope. Sharing can happen in a casual sort of way, with one person lending another this or that. Or it can be more formal, with two or more people having common ownership – the arrangement we have with chain saw and mower. However you organise it, the benefits are fairly obvious, and, apart from the savings in outlay you can make, you will also make new friends. Our neighbour Cliff, for example, has been exceptionally kind to us in lending things, ranging from tractor and trailer for carting hay, to hay knife for cutting it out of the stack, and a fine old traditional 'fiddle' for sowing seed. So far, all we have been able to do in return is slip him a small bag of potatoes when his crop had failed. But our time will come.

Inevitably sharing has its disadvantages too. Some people cannot be trusted with their own machinery, equipment and tools, let alone someone else's, and mutual discomfort is inevitable when you discover excess wear or damage to a returned item. You will learn by experience – as we have done – but one way to avoid undue pain is to 'try out' untested borrowers with something small or non-damage-prone before you lend them anything expensive and delicate. apart from this, things may not always be available when it is your turn to borrow; and when you want something of yours returned, the borrower may not be around, or he may be right in the middle of a job and you do not have the heart to interrupt him. All the same, you will probably find, as we have, that sharing is a nice way of doing things when you can. Finally when you can run a 'transport' pool so well that you escape from the extravagence of 'one family one vehicle', then you will know that you have arrived!

Avoiding waste

Shirley and I have found that there are two principal ways to avoid waste: by not buying machinery and equipment unnecessarily; and by caring for what we have. And the caring takes two forms:

maintaining things so that they last longer; and storing them to keep out the elements.

I am not very good at maintaining things, and I suspect I am not alone in this defect. When I get my new piece of machinery home, I am like a kid with a new toy. I fill it with fuel or whatever, read the manual to see which levers to push and pull - and *use* it. I am not a complete fool, mind you. I know that things need oiling from time to time, and so I look at the drawing in the manual which shows the average idiot where to push in the oil can - I do so and then back I go to the job. For years the detailed maintenance instructions have been to me like the Small Print in an insurance policy - not to be read for fear it says something difficult or otherwise unpleasant. I think I have now broken through the maintainance barrier, but only after many a chastening experience.

The sadness of the rotovator is one of them. To attach the trailer I must remove the tynes. Simple. And almost as simple to put them on again. Except that the manual says they must go back a special way, and that if they do not the transmission will be 'subjected to undue strain'. I missed that part of the manual and now my tynes have a disturbing wobble. Then there was the sadness of the chain saw. The manual says 'clean the air filter regularly'. Impatient to keep sawing, I did not. Result: air intake clogged; engine lost power and oiled up, chain saw cut at too slow a speed and blunted the chain; time and fuel were wasted; engine had to be decarbonised; chain needed premature sharpening, and - worst of all - relations with my co-owner were strained.

Now the moral of all this is to remember always that machinery deserves to be cared for - to be oiled and greased, to have its bolts tightened, filters changed, tyre pressures topped up, its outside protected from rust - and so on. It is the same with equipment. Garden tools especially need Tender Loving Care, for too often, after a tiring day, any one of us is likely to hang them up muddy and wet, and so encourage rust. We keep a large tin of sand mixed with sump oil handy, and the rule is to scrape them free of mud, then push them up and down in the tin until they come out clean. To neglect this sort of caring with machinery and tools is not only to invite trouble and expense, but to be guilty of incurring unforgiveable waste.

I have not been very good at storing things either - not only machinery and equipment, but fuel, fertilizer, wood, hay and harvested crops, and since the point I am making concerns waste in general, and not just waste associated with machinery, I shall tell a broad spectrum of chastening sadnesses. You will probably find, as we have, that certain tasks *defy* you. They simply do not get done. They lack charisma or something, for other tasks present themselves more forcibly and get done in preference. The longer these defiant jobs remain undone, the harder it becomes to get around to doing them. They become a permanent feature of the place. We cannot explain this phenomenon, unless it is that we are less than perfect ourselves and occasionally procrastinate.

The trailer is a case in point. When we bought it I realised we had no suitable shed to store it in, so it had to stay outside. It was supposed to be made of waterproof plywood anyway and so we did not worry. Rain came and stayed for weeks; burning sun came and lasted for months, and the trailer defied us by staying where it was. About a year ago I hit on the idea of covering it 'temporarily' with a couple of odd sheets of corrugated asbestos, which protected it, but not before the elements had permanently warped its wooden base. This experience demonstrated how little faith one should have in manufacturers' claims, and constantly reproached us for our gullibility and procrastination. And it forced us to act. Shirley cleaned it and applied two good coats of marine polyurethane sealer, which saved it from further warping, and I eventually built a shed for it.

The importance of good storage can hardly be over-stated. That you have inadequate buildings, as we had and still have, must not deflect you from building what you need. Hay demands a dutch barn. We do not have one, so we made stacks of loose hay and threw polythene sheets over them, weighted down with bricks. The wind soon ripped the sheets so that the rain got in; and where it did not we still lost hundredweights of hay through dampness from condensation - for plastic traps rising moisture as hay cures. Now the experience has stung us sharply enough to make fresh storage space for all the timber and lumber cluttering our one and only barn, so that we can store our hay there - as we should have done in the beginning. Firewood has to be kept dry. We covered our stack with plastic, but it blew away, ripped and sagged so that rain still soaked it. Now we have built a special shed for it.

For us, lime has been a sticky problem - literally. The trouble is that if lime gets wet it sticks together in cakes and is ten times harder to spread. When our load of lime was due to be delivered - several tons of it in paper bags - our neighbour, Owen, kindly offered to let us store it in his big shed. We accepted, in due course unloaded - and forgot about it. Several weeks later when spreading time was imminent I went up to his place. We had had heavy rain, and I wanted to see if the ground was firm enough to take my van when we came to load. The first thing I noticed was that the shed had disappeared. The second thing to hit me was that it had blown down. And the third was the state of our lime - soaked through and through. Owen had thrown a plastic sheet over it - but like the shed - the plastic had promptly blown away! We did eventually get most of the lime out, though we have no wish to repeat the experience.

I could go on. Inadequate storage wastes more than *things* however; it wastes that other valuable finite resource - *time*. We bought half a dozen old petrol cans in a sale and lost quite a lot of petrol through tiny leaks caused by rust and corrosion. We thought our troubles were over until the rovotovator kept stopping. I took the carburettor off and found minute specks in the float chamber which were blocking the fuel intake. Even though I cleaned the carburettor thoroughly and eventually took off the fuel tank and cleaned that too, the trouble persisted long after I had discarded the last wretched can.

Inadequate, untidy storage - such as we still have - wastes time in another way. You can never find what you want! You are almost set to start a job, but you still need the bolster chisel - counter-sink, bone meal, wire strainer, soldering flux. You have put the very thing you want in some 'safe place' for this occasion ... but you cannot remember just where. I wish I had ten pence for every five-minute to half-hour delay caused by the perversity of things to hide themselves just when they are wanted. And it is not as if the delay concerned just one of us. Invariably the other must down tools to join in this game of hide and seek. Only one thing eases the pain: we have discovered we are not unique.

CHAPTER ELEVEN

The Lessons

What might be learned from the foregoing pages?

Before embarking on your venture you have plenty of homework to do. You can prepare for it generally by weaning yourself off city habits, such as convenience foods, eating out, entertainment and travel; you can learn about farming and other useful skills, meet like-minded people, read; and acquire capital by saving, sharing, working and being enterprising. You can buy tools and other essentials while shops are handy. You can examine your motives for moving and distinguish the positive ones such as a search for self-reliance from the negative ones such as escapism - distinguish fact from fancy. Among the illusions of rural self-reliance are the ideas that the life is all tranquility and bliss with little need to organise nor even to work especially hard, that food is free, and that it is a simple life. The realities include streams of hungry visitors, less contact with other self-supporters than you probably imagined and more difficulty adjusting to country life and ways - the joys of country living may not be readily apparent.

You may find after all that you are unsuited to the demands of the life, that you are contemplating it for the wrong reasons, and that you really lack the necessary resources of energy, determination and money. In which case consider some other venture. If you possess even some of the resources, however, and you carry on, you have the prospect of a way of life perhaps unequalled in its rewards. But even so you must learn to live with insecurity.

Before you settle, draw up a rough economic plan of how you propose to feed yourself and provide for your other needs both from your land and beyond it, though do not expect cast-iron figures of costs and monetary returns to guide you. Your land may grow your food and fuel and you may make your own clothes and candles, but you will still have bills to pay. And so the importance of regular income cannot be over-stressed; even though you may need less than half the city norm, it could prove twice as hard to earn where you are. You will discover three basic constraints in all you attempt - time, stamina and money. Do not expect to be fully self-sufficient in food or other needs. Unless you are 'cheating' with an unearned income, plan for an income from a craft or other work, but not from a full-time job unless you see yourself on only a tiny patch of land. The more time you spend earning, the less you will have for farming. Decide whether you want a large garden or a small farm, being guided by how you see yourself - as gardener or farmer.

Avoid attempting too much at once, for every project has its 'down payment' and 'instalments'. Total self-sufficiency makes no sense; the closer you strive to be, the more complex and exacting your work becomes, and the more ascetic your life style must be. To practise self-reliance successfully you need to be an unusual person. You will need some brains, imagination (five-year vision), energy, determination, self-discipline, resilience, resourcefulness, a sense of humour, organising ability, an appreciation of priorities, friendliness, tolerance, humility and an interest in nature. Physical fitness is a 'must'. Your partners must be convinced of what they are doing and relationships well-founded.

Again before you settle, decide on likely districts and explore them. Consider latitude, altitude, rainfall, how windy, land prices, soil fertility, scenery, proximity to income possibilities and like-minded people. Avoid making up your mind on a sunny day. Check on the chances of industrial and road developments. Talk to as many as you can when visiting. When you have found your district, look for your place. Avoid too much steep, poor or wet land, and frost pockets. Get the soil tested. Avoid any derelict place unless you have the energy and resources to develop it. Be seduced neither by the view nor by an old house with 'possiblities'. You will need outbuildings, and if they are not there you must build them. Talk to people around about any place you fancy. When you have decided on your place, sharpen your economic plan accordingly. Draw up a long-term plan - say five years - based on the economic plan to get priorities right and for a time scale and costs. Alter it at any time according to feedback, and produce annual plans to work to.

Make lists of work to do and keep essential records. Calculate how much of each crop to plant, being generous to allow for failures. Start your days early and keep some structure in them. Take a regular day off each week – 'Special Day'. Discard props and crutches one at a time. Accept setbacks and suffering, but remember that they enrich the rewards. Do not farm or garden until your house is habitable and any conversion work done. Do as much as you can yourself. Make sure any 'experts' – architects and builders and so on – are qualified. If you want to play safe, and can afford it, employ a proper builder, but add 50 per cent to estimates and have enough funds to cover. Try not to make 'improvements' and other changes during building. Make your house match your intended life style. Go easy on alternative technology. Aim for a large kitchen and ample storage space. Think twice before deep-freezing, and minimise kitchen gadgets. Make bread without kneading. To stay healthy eat little or no meat, but cheese and eggs instead. Use imagination in cooking vegetarian or low meat meals.

Understand your land and treat it kindly. Remember the vital chain of birth, death, decay and rebirth. Grow crops organically, using chemical fertilizers only for exceptional circumstances. Do not leave the soil bare, nor work when too wet, nor overgraze pasture. Try not to be too greedy, but accept some losses to birds and bugs – they have to live too. Do not try growing unsuitable crops, but instead co-operate with other self-supporters who can grow what you cannot. Work out a crop rotation and keep to it. Use heavy machinery as a short cut to development if you are not-so-young.

You can neither insert yourself into a country community nor buy your way in: you must be invited. You have to earn respect – by work and by repaying kindness with kindness. Do not act superior or patronising, and remember that people are people first, functions second. The future will always be uncertain, but if you are doing what is right, people and resources will form 'The Flow' which will help you through hard times. People will come to you in greater numbers than you expect, to help and just to see. You will bless some and learn how to cope with others. Bureaucrats will also visit you, some friendly, others less so, and it is best to avoid confrontation. The people that matter most are those sharing your venture, so discuss it fully before you move and afterwards – children especially for they will be much affected. Encourage adventure for them and discourage TV. Insist they do chores and reward them. Learn the joys of homemade fun and entertainment.

Livestock are a tie, for they mean constant work and occasional trouble, but on balance they can be good for you and your land. You must decide whether or not to keep them; if you do, you will want to keep cows or goats - or both. Consider the merits of each. Goats suit rough land, and if properly fed and fenced need not wreck your garden. Cows are quieter and their milk is more versatile, but they need good pasture and have huge appetities. Seek advice when buying either. If you keep more than one cow you risk an encounter with officialdom for you can sell goat's milk and products with little restriction, but not cow's milk and products. Cope with milk surpluses by making cheese and butter, planning beforehand. For your protein needs, you may do better eating surplus milk as cheese, giving the whey to pigs and poultry, rather than rearing kids and calves on it. Unless you keep pigs and poultry only on a small scale or grow your own feed, you may find that bought-in feed renders them unprofitable.

Remember that you do not live by food alone - so learn other skills and use them. Much farm work is moving things around, but keep machinery minimal. Call in contractors or neighbours with machinery for heavy work. For a small farm run a secondhand tractor, for a large garden run a rotovator, if steep land, with powered wheels. Run a pick-up or van rather than a car. Improvise, share and buy secondhand where possible. Take care of tools and equipment. Store everything properly to avoid damage and losses. Everything is constantly wearing out and distingegrating; success largely depends on being able to keep things running and together at a faster rate than coming apart.

For your venture you will need capital and income, and there are ways of acquiring both. You can manage on less if you pool resources with others and live more or less communally. Do not saddle yourself with a mortgage or other large or long term debts. To reduce income needs, live simply and co-operate with others. Neither let your income earning interfere with your land work nor compromise your ideals.

The wider relevance

Living on a little land, though precariously dependent on the personal qualities of the practitioner and on a source of income, can clearly offer a rewarding alternative to the city 'rat race'. This may be attractive to the individual, but has it much relevance to others? Practised properly, we believe it can. It is a way of life which takes *less* from industrial society and the planet's limited resources, and gives back *more* in the form of food and work directly beneficial to ordinary people. These resources comprise not only artifacts and services produced by city people commited to boring, unrewarding work, but also the prime essential of energy – the safe non-renewable fuel used in making and transporting artifacts and chemical fertilizers. Because the farming methods are organic, the land is not 'mined' – it stays fertile.

It is a way of life which helps to bridge the widening gulf that separates city from countryside, for it attracts helpers and observers who are not yet able to farm their own land. All these people are able to learn, participate in and enjoy – not only food growing but the whole spectrum of self-reliance; and in overcoming their helplessness they begin to discover themselves.

It helps to create a less centralised society. Already it is reviving a countryside sick from a surfeit of commuters, retired couples, weekenders and vacationers, and a dearth of young, energetic workers – all of whom are far more concerned with the health of their land than their balance sheets. It is an intensive system of farming which, with less fuel input, produces more food from each acre than comes off each acre of vast agribusiness holdings. Since the average size of farms is continually growing, the movement offers a stalwart resistance to the general trend. Intensive farming could reduce the nation's necessity to import food from abroad where, because it is needed to feed hungry people, its supply cannot be guaranteed.

The movement can bring into production countless acres of marginal land now scorned by big landowners. Such land can only realise its potential by the kind of money, love and detailed attention which emanates from people hungry for land and simultaneously concerned with the need to grow more food. It is an inspiring model – already working – of a joyful, alternative way to live which only requires reforms in land tenure laws to become a powerful force in the land.

Living on a little land can produce an ample surplus of nutritious and delicious food, free from the poisonous taints of pesticides and herbicides. All these are powerful arguments, yet there is still one more that must be stated.

The message of the land

You cannot become part of the country overnight, nor even in the fullness of a whole year. For still longer you may live in the country, revel in growing food and husbanding stock, delight at the variety of wild flowers and trees around you, enjoy the presence of wildlife, and find companionship with country people. And remain an outsider.

When you can stand in your garden on a still spring day and *feel* the life around you, opening, swelling, blossoming, vibrating; when you can sit as still as a stone at evening in dark green woodland and hear the teeming life around you, the high, tiny whine and ping of insects, throbbing of wings, answering cries and padding of soft feet, all at one with the pounding of your heart; when you can smell approaching rain and read the future in the clouds; when you can sense that you are neither greater nor less than any of these, you are drawing closer to the essence of the country.

Past generations of country people could hear and read the language of the country for they were born into it; and that which they did not absorb as children was made known to them when they were ready to receive it. The wisdom which accrued to them was both echo and reflection of an intelligence set apart from the built world of man. It concerned the mystery that made plants and trees grow, instincts of animals and birds, secrets of the soil. Like all mankind the countryman had separated himself from this intelligence if he were observer, poet or artist enough, but he might glimpse it; and at rare times he might even be tuned finely enough to become part of it.

The new generation of country people is not so privileged. As children they could have heeded the same wisdom, but instead they were bombarded with city-centred television, and the schools which replaced the village schools were similarly dominated. If any of them continued to agricultural college or university their teaching was influenced by the immense power of the fertilizer and drug companies. If they opened the pages of farming magazines they were confronted by page after page advertising poisons with which to dose their land and their stock. Knowledge of the ancient wisdom of nature has been steadily supplanted by the 'technical fixes' of science.

Few true country people remain. Many people still live in the country, for the car and the motorway have made it a playground from which to commute to towns and cities, and a relatively quiet place for retirement. Despite the exodus of farm workers to towns and cities, there are still some left on the big, business farms which have largely absorbed smaller family farms. And there are still men and women who cling to their birthright farms. But too many farm workers spend too many days in a tractor cab and never touch the soil; or mechanically feed and tend stock in factory 'farms' where there is no soil at all. And the remaining farmers are so deep in debt to financiers, so dependent on chemical fertilizers and pesticides, and so alarmed at escalating fuel bills for their armament of machinery, that they are more worried about the health of their balance sheets than the health of their land. Their first love is money. Love of the land as a way of life is fast vanishing – if indeed it has not already gone. Crops and stock are now a means with which to pay bills; the soil is merely a medium to prop up the crop until it is ready to be cut down; and a surface barely firm enough to take the weight of the machines which will trundle in to do the job.

Already a third of the farms in Britain are engaged in the monoculture which thrives upon such attitudes and practices, and the proportion is constantly growing. And since the land is now recognised as one of the few safe repositories for investment and pension funds, large companies and speculators are buying land and farms as fast as suitable properties become available.

In the remoter districts beyond commuter belts, in hilly country and regions where land tracts suitable for 'development' are scarce, the traditional, smaller farmer is making his last stand; but even he must fight to evade the salesmanship of the poison peddlers and the clutches of the financiers.

This is the dubious scene that the self-supporter enters when he leaves the city to live out his dream. His affinity with commuters and with retirement couples is slender – if any exists at all. With the 'agribusiness' farmers and their managers he has none whatever – unless he happened to go to the same school. Initially the smaller, traditional farmer will view him with suspicion and a good deal of not unfriendly amusement. If there are other self-supporters handy, they are likely to be over-engrossed with their own problems and, unless they are almost next door the newcomer cannot expect to see them very often.

Relationships with other self-supporters are to be cherished: they will need each other and no one knows how soon or how desperate the need will be when it comes. Any opportunity to draw close to a true countryman or woman is to be welcomed and pursued, for then self-supporters may glimpse wisdom that is all but gone. For much of his time, however, the self-supporter is on his own or alone with his family. His thoughts are his constant companions. His land will dominate his thoughts.

He will reflect on the past and sometimes wonder why he came, or why to this special patch of his; he will immerse himself in the present, preoccupied with solving immediate problems; and he will speculate on the future, wondering what the seasons will bring and whether his craft or other work will be lucrative enough to meet his needs. Yet if he is wise or fortunate, he will find his thoughts roaming through other pastures. The wisdom of the country which is all but lost is nevertheless not locked away. Anyone who works to become part of the country and who is humble enough to discard his role of 'owner' can discard the rubbish from his mind and heart and relearn what he has forgotten. A sense of who he is and why he is here, of the beauty and order of nature and the authority of her laws, of a force immensely greater than himself - beyond and yet within him - which some call God and which invites total trust - all this was there once, long ago when he was a child, and countless years before that time. Anyone who can rediscover even a morsel of that wisdom will find his work lighter and his harvests more fruitful. He will spend less time fighting his seeming adversaries, more time getting to understand them. He will expend less energy wrestling with the soil, for the soil will share its abundant energy with him. He cannot expect an easy time, for there is nothing *natural* in comfort and security, as any study of the wild will quickly reveal. Work there will be, as there has always been; but work will take on a different feel until the word no longer has its old meaning. Over a period of time - and for each of us its length will vary - he will come to feel less *separate* from the country, less of a visitor. The sense of being a *participant* will slowly develop, and as it does, so will a sense of kinship with the other creatures which share his land, and with the few people around him who still remember the old wisdom.

Those of us who discover in this way how to participate will gain a rare honour - the inheritance of the old wisdom and the responsibility of being the new generation of country people to cherish it. So that when the cities can no longer be sustained and people come to their senses, those of us who have ventured forth to return close to the land will have much work to do. From us and others will the fugitives be obliged to understand the land, blending the old wisdom with as much of their own as is appropriate for the health of all. The task will take time, however for many wounds must heal. And you cannot become part of the country overnight, nor even in the fullness of a whole year.

SOURCE GUIDE

How and where you can learn more and meet people. Just a few of the organisations and publications in the field - but, except for books, without fees and rates because they so quickly go out-of-date.

ORGANISATIONS

COMBINED ORGANIC MOVEMENT FOR EDUCATION AND TRAINING: COMET for short, it organises day and weekend courses in organic growing. Stamped addressed sticky labels to Dick Kitto, Lower Shaw Farm, Swindon, Wilts, SN5 9PJ.

ECOLOGICAL LIFE STYLE LTD: provides a way for people to change gradually from city life to greater self-sufficiency by subscribing to the Ecological Land Bond Fund. Money so raised buys farm land with which bondholders become increasingly associated. (See page 131). Newsletter published. 27 Burnham Road, St Albans, Herts.

FRIENDS OF THE EARTH: besides working for a responsible attitude to the environment, runs Crops and Shares Scheme to help landless people grow food. 9 Poland Street, London, W1V 3DG.

HENRY DOUBLEDAY RESEARCH ASSOCIATION: encourages organic horticulture, conducts field trials and publishes quarterly newsletter. 20 Convent Lane, Bocking, Braintree, Essex.

RURAL RESETTLEMENT GROUP: aims to stimulate and support a movement to the countryside of people wishing to live and work in sustainable rural communities on principles of co-operation. Information handbook available. 60p plus SAE 12½p to Lower Shaw Farmhouse, Swindon, Wilts, SN5 9PJ.

SKILLS EXCHANGE NETWORK FOR A SUSTAINABLE ECONOMY: SENSE for short, it compiles lists of people willing to teach others their skills to encourage self-reliance. Free to those offering skills, 14p in stamps for those wanting to learn. News bulletins for SAE. John Porter, SENSE, The Forum, Chidham Park, Havant, Hants.

SOIL ASSOCIATION: promotes soil fertility through organic growing methods. Local branches. Publishes journal. Walnut Tree Manor, Haughly, Stowmarket, Suffolk, IP14 3RS.

WORKING WEEKENDS ON ORGANIC FARMS: WWOOF for short, it arranges instructive work on farms in exchange for food and lodging. (See page 108). SAE for details to Don Pynches, 19 Bradford Road, Lewes, Sussex, BN7 IRB

LITERATURE

Communes Network: newsletter of the movement. Monthly. Lauriestone Hall, Castle Douglas, Kirkudbrightshire.

Bring Me My Bow: by John Seymour, spirited advocacy for revitalising the countryside, reforming land tenure and general self-reliance. Paperback £1.95. Turnstone.

The Complete Book of Self-Sufficiency: by John Seymour, as complete as possible in 250 pages, £5.50. Enlarged version of earlier *Self-Sufficiency*, £2.95. Both hardbacks, Faber.

Food Growing Without Poisons: by Meta Strandberg, introductory handbook to organic gardening. Paperback £1.50. Turnstone.

Living Better on Less: by Patrick Rivers, a philosophy of self-reliance which makes a low-impact life style enjoyable. Practical advice on diet, clothing, housing and other chief areas of possible savings. Paperback £1.95. Turnstone.

Practical Self-Sufficiency: what its name implies. Bi-monthly. Broad Leys Publishing Co., Widdington, Essex, CB11 35P.

Technological Self-Sufficiency: by Robin Clarke, shows how to learn basic building skills and alternative technology applications. Paperback £2.95. Faber.

Vole: environmental and literary magazine covering revitalising of the countryside, and follies of the State and big business. Monthly. 20 Fitzroy Square, London, W1.

Whole Earth: newsy magazine on self-sufficiency, organic growing and alternative technology. 11 George Street, Brighton, Sussex, BN2 1RH.

THURS 31st OCT.

WED 6th → 23/24 NOV.